"What kinds of earth changes must we expect?"

"The seers and the scientists, we shall discover, are in agreement. The avenues along which the radical transformations of the earth will travel are many. They include increasing frequencies of earthquakes and volcanoes. They include trends toward a global cooling spell arising from atmospheric paroxysms of adverse climate conditions. The higher incidence of earthquakes and volcanoes will shake us to the core. Changing weather patterns and climates will make our very survival a constant struggle.

"... But this book is not a prophecy of doom. Its purpose is to make you, the reader, aware of all the components that weave together to create our earth's condition. The more we learn about the earth and how man fits into his environment, the better we can prepare to meet the unexpected."

...BUT WE MUST ACT NOW!

EARTH CHANGES AHEAD

FRANK DON

Authors Choice Press

San Jose New York Lincoln Shanghai

Earth Changes Ahead
The Coming Great Catastrophes

Authors Choice Press
an imprint of iUniverse.com, Inc.

For information address:
iUniverse.com, Inc.
5220 S 16th, Ste. 200
Lincoln, NE 68512
www.iuniverse.com

Originally published by Warner Destiny

ISBN: 0-595-18686-6

Printed in the United States of America

DEDICATION

To the spirit of universal brotherhood.
By joining hands in cooperative effort,
humanity can survive the earth changes ahead
and create a paradise on earth for all.

ACKNOWLEDGEMENTS

I wish to acknowledge the contribution of the following people in the research, writing and publication of this book.

Dianne Flynn
Kendall Gardiner
Dr. Henry Grinberg
Bill Habekost
Nancy Hoffman
Bob Jackson
Joanna Jackson
Alden Cole
Dianne Metzger
David Pearlstone
Bob Scarpulla
Marie Scarpulla
David Singer
Ehud Sperling
Lisa Sperling
Sheila Steckel
Susan Resnik
and the scientists who have contributed their lives
to expanding man's understanding of his earth
and his environment.

CONTENTS

INTRODUCTION

IS THE SKY FALLING?

Make no mistake about it. There are earth changes ahead. Of course, we don't like to think so. This solid old earth of ours change? That's impossible. How can it? How could it?

When we were children, we heard the story of Chicken Little. While she was outside one day, an acorn fell on her head. Thinking the sky was falling, Chicken Little set off to warn everyone she met. But all the folks were wise. No, the sky was not falling, and we were reassured. For a hesitant moment, we ourselves might have gone out of doors, staring up at the infinite blue, and realized that the sky had not fallen— and would not. And we covered ourselves with a veil of security— or was it complacency?

Recent catastrophic earth events are pieces of a voiced cry emanating from the depths of the planet. The volcanic eruption of Mount St. Helens in Washington State was long ago prophesied for a time such as ours when the gaps between earth humanity and its sacred home had grown so wide as to jeopardize all life, as to call the future itself into question. Were mythic journeymen to venture today into the belly of the earth, their reports would be alarming. Blaring sirens would be heard sounding the warning to prepare for a time of trial ahead. In her own conscious way, the Earth is about to embark upon a last-ditch effort to awaken us to our collective irresponsibility, to call us on our behavior as a species run amok, unaware of the limits of our biosphere's regenerative capacities.

In making the changes we know to be necessary for our time, we must make them total. Economic, political, or social reconfigurations must be grounded in physical reality. An expanded awareness and reach of the human mind, altering as it does the very definition of ourselves and our planet, implies a physical alteration to come one day. Becoming aware of how the in-breathing and out-breathing of earth's being has accompanied the birth and death of countless species, communities, and civilizations will help us to prepare for the earth changes ahead.

A rash of books has appeared in the last decade, all speaking out for Her, the Earth Mother. Each has sought to call our attention to the urgency of Her needs, to the difficulties and sufferings She has borne for us during this latest industrial phase of our social evolution. Some have tried to awaken us with shock treatment, portraying for us in prophecy or in science fiction the havoc that man's carelessness can wreak. Others have revived ancient traditions of prophecy, bringing them to the attention of modern publics. In their particular ways, each of these books has raised an aspect of the larger dimension which *Earth Changes Ahead* attempts to encompass. The earth has been warning us in many ways that we are upsetting the natural functions of our planet's operations, and yet we seem to remain ignorant, to shut our eyes and ears to the truth that surrounds us daily. The living species that inhabit the land and oceans are vital to the very air we breath, the food we eat, and the water we drink. Everything is interrelated. It's time to face the reality that our existence as individuals, as families, as nations, and as a species is intimately tied up with and dependent upon the earth's ability to progress on its own terms. We are out of synch with the obvious. Are we in the process of acting out an immense tragedy which may lead not only to our own destruction but to that o the whole world? Can we avoid it?

This book is not a prophecy of doom. Its purpose is to make you, the reader, aware of all the components that weave together to create our earth's condition. The more we learn about the earth and how man fits into his environment, the better we can prepare to meet the unexpected. We shall discover that vital changes result not only from natural events but also from man's actions as well. For too long we've been ignorant **of the laws that govern our earth.** For too long we've considered the earth just another artifact for us to manipulate at whim. For too long we've thought that events are isolated, that individuals and governments might abuse the earth and its atmosphere without penalty.

Our illusions about a "solid," unchangeable earth have been blasted. We know now that there is no such thing as "terra firma." Earth is not a lifeless, rigid, and absolute mass. The scientific community increasingly recognizes the earth as a dynamic, living entity. We now realize that **our earth has undergone radical changes in the past.** Continents have split apart. **Mountains have pushed up from the ocean depths.** Great land areas have disappeared beneath the seas. The forces that cause continents to drift continue to do so today.

We are all planetary citizens who can and must mitigate the effects of past actions and must anticipate radical alterations to the earth in the days ahead. Not only do such changes result from internal pressure and shifts of land mass, but they also seem to occur in conjunction with certain cosmic events. We shall investigate these interrelationships, and the earth's internal dynamics, to understand the effects we experience as drastic change. While phenomena such as earthquakes, *tsunamis* and volcanoes leave man shaking with impotence in the face of their awesome force, the knowledge of why and how they occur can help dispel fear and enable us to be prepared for such catastrophes.

What is man's role in precipitating some of these disasters? Through his unthinking desire to "better" his life, man has altered environmental conditions according to his own dictates. When he decides to improve nature, he has often done so with little or no thought as to the long-term effects of these changes. Weather pattern changes, for example, are a reality in nature. Man's tampering with weather processes has wrought changes that have upset vital life-support systems.

What kinds of earth changes must we expect? The seers and the scientists, we shall discover, are in agreement. The avenues along which the radical earth transformations will pass are many. Earthquakes and volcanoes will occur with increasing frequency. Trends toward a global cooling spell will arise from atmospheric paroxysms of adverse climate conditions. A higher incidence of earthquakes and volcanoes will shake us to the core. Changing weather patterns and climates will make survival a constant struggle. The chapters of this book will chronicle earth changes past and present, exploring the ways in which they have been explained, as well as indicating what we should expect in the years ahead.

The present era of the earth's history has been relatively quiescent. Nonetheless, there are approximately 500 live, active volcanoes. More than one million earth tremors occur each year, with an annual average of nineteen considered major earthquakes. Earth spasms, whether volcanoes or quakes, generally occur along the margins of dynamically mobile lithospheric plates which move apart or override one another. Two major belts of volcanic/earthquake activity girdle the globe, but earth spasms can hit unexpected areas, too. Volcanic eruptions produce not only cataclysmic local effects, but can cause crises of hunger and famines in other areas of the world because of temperature drops that turn summers into winters and ruin agricultural harvests. Earthquakes can generate such assorted effects as landslides, avalanches, *tsunamis,* and fire storms.

How can we explain these earth spasms? We'll consider several theories dealing with the relationships between earthquake activity and other natural phenomena. These theories are subject to dispute within the scientific community, but most scientists acknowledge that the current quiescent geological phase will give way to increased earthquake and volcanic activity in the future. The correlation between earthquakes and the coincidence of two cycles associated with the earth's global wobbling signal a recurring threat of increased earthquake activity every seven years, which in future years would occur in 1985, 1992, and 1999. We'll see how the approaching alignment of planets, scheduled to occur between 1982 and 1984, is likely to trigger earthquake activity in accumulated strain areas such as the Los Angeles and San Francisco regions of the San Andreas fault in California, and on the island of Honshu, where possible death tolls could run as high as three million.

As we move into a period of increased seismic activity,

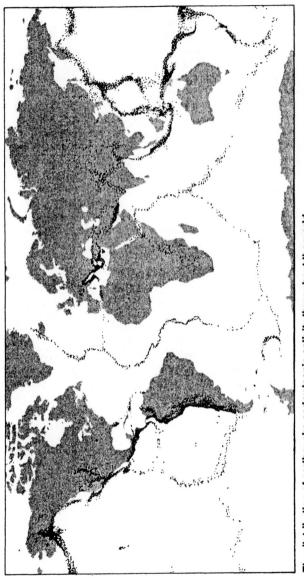

The distribution of earthquake and volcanic activity throughout the world.

we should be aware that no area can be considered totally safe from earthquakes. When the earth writhes and releases its pent-up energies, unknown dormant fault lines may spring to life with explosive force and catastrophic consequences. Take, for example, the intraplate earthquakes that hit Charleston, South Carolina, and New Madrid, Missouri, during the nineteenth century.

What can we expect as a result of increased geophysical activity in the future? Volcanic eruptions inject greater amounts of particulate matter, or dust, into the atmosphere. From now on, the increase of this particulate matter will accelerate even faster, due to the increasing rate of man-made pollutants. Higher dust levels promote an aerosol veil that will significantly reduce the amount of solar radiation reaching the earth's surface. The resulting change in the balance between incoming solar radiation and outgoing terrestrial radiation will also cause increased global cloud cover and effect land use, altering the reflective qualities of the earth's surface. There are climate changes ahead.

The chapters devoted to the earth's climate will examine the natural cooling trend of the present era which, provoked by geophysical variations and by man's actions, will become more pronounced and lead to swift changes in future climate conditions. The current debate over whether the earth is moving toward a warming trend, a cooling trend, or no detectable change, may be resolved by the mid-1980's. Evidence points in the direction of a significant cooling trend which will weaken atmospheric circulation, creating more pressure centers of a smaller and less potent variety. These weak pressure centers tend to stagnate over areas for long periods of time, producing unseasonable cold or warm weather to one side of the center, with reverse conditions on the other side. Areas affected by a warming phase will experience extensive flooding. Reduced evaporation and subsequent reduction in precipitation make drought the greatest threat to the world population during a cooling phase. In high latitudes, the expanding snow cover will make cultivation of the land impossible. In moderately high latitudes, the agricultural growing season will be shortened. The decline of soil fertility and the falling rates of annual food production will deplete food reserves, and eventually cause starvation.

In the twentieth century, mankind has enjoyed a climatic optimum unparalleled since the eleventh century, but the bountiful agricultural conditions that promised to eradicate world hunger are changing. During this relatively plenteous period of human history, over 455 million people are suffering from chronic malnutrition. Within 25 years of a sharp cooling trend, malnutrition will degenerate into starvation, and may claim as many as 500 million lives.

How will man react to the terror and the agony of a future that brings sharply colder climate conditions and accelerating frequency in earthquakes and volcanoes? Perhaps he will cry out in anguish for an understanding of the living system from which he is so firmly divorced. But who will hear him in a world he has deemed an existential void? Perceiving loneliness in the universe, man will be hard put to maintain any semblance of mental balance. If the prophecies prove correct, it seems that man's initial response will be instinctual. If his life supports are threatened and his inner strength is weak, man might be reduced to barbaric means of survival. Aggression, war, and anarchy are likely by-products of man's fall from civility. Who can deny mankind's temptation to misuse knowledge, as material standards of living drop and the lack of food threatens human survival?

The legend of Atlantis provides us with an interesting parallel to the conditions that modern man may soon be faced with. The Atlanteans employed a knowledge of natural forces— a knowledge even further advanced than our own twentieth century wizardry—in behalf of greed and negative purposes. Their misuse of knowledge and power led to the destruction of their glorious civilization. Perhaps modern man will do the same. Tampering with intense natural forces, he has developed the technologies necessary to engage in climatic, geophysical, and environmental warfare. We have been warned. Thanks to the seers and prophets of the past, who were able to detach themselves from worldly circumstances to take a larger view of life, we can gain additional insights into man's unfolding history, and modify our outlook for the present and for the future.

The opening chapters of this book will explore various prophecies which have shed some light on past and more recent earth events. We recognize, however, that prophecies cannot be validated until after the fact. It's unwise for us to accept prophetic statements blindly. The possibility of error that lies within us is not to be ignored. Nor can we ignore man's ability to change his destiny by choice and free will. Nonetheless, the heightened sensitivity and consequent prophetic insights of persons such as Edgar Cayce have left an indelible imprint upon man's consciousness. A full chapter will be devoted to Cayce's unique understanding of earth mechanics. The remarkable legacy of Cayce is still in the process of being validated by world events and scientific discoveries.

We'll also consider the interpretations of astrological knowledge which for centuries have been hidden away as superstitious folly in a dark corner of man's consciousness. Astrology, however, does provide another piece in the puzzle of future earth changes. Through the mechanics of astrology, man seeks to understand his destiny. Whether the planets actually affect man's destiny, or merely reveal it to his understanding, much as a clock reveals the time of day but does not cause the hour, is a theoretical problem which is not relevant to our subject. What is relevant, however, is astrology's role in revealing world destiny, a basic tenet which dates as far back as the days of Chaldea, when astrology was used exclusively for this purpose. After reviewing what Cayce and astrology have to say about future conditions, we shall discover how scientific research echoes a similar message: Earth Changes Ahead.

We would like to lead you on a walking tour of our dynamic, living earth, to help you to see for yourself the changes on its horizon. We will be guided at all times by scientific reason and research, but we have also decided to take along two imaginary characters who will visualize more concretely for us the dialectic of the struggle within the understanding of change. We hope thereby to stress that beyond the apparent conflicts, there exists a positive and productive relationship between scientific analysis and imaginative synthesis, something akin to what the poet William Carlos Williams meant when he referred to the discoveries of Mme. and Pierre Curie:

A dissonance
in the valence of Uranium
led to the discovery

18

Dissonance
(if you're interested)
leads to discovery

Our imaginary characters will be designated simply by alphabetic symbols (life chemical elements): L and R. We invite you to imagine proper names as you will or to keep them abstract: the left and right poles of dialectic localized in a mind struggling with itself, or even two feet alternating their movements. In any case, the result will be the same—a shifting of balances which moves through periods of imbalance. Call it thinking or walking, it is still forward movement, progress.

The immediate future may test humanity with radical physical disruptions of our earth. Mankind's level of cultural evolution, as well as his spiritual evolution, manifesting as *good will*, must challenge the competitive ways of the past. It is the author's belief that man can use these challenges as a catalyst to transform his consciousness. Do please join me as we move and think toward the future, facing the Earth Changes Ahead.

I

SCIENTIFIC AND PROPHETIC VISION: TRANSFORMERS OF THE EARTH

R: I don't believe it. I don't believe you can predict the future. It's simply a matter of common sense. We can only know what *is*, what we examine and see, now. The future is the future; it does not yet exist. Therefore, it cannot be known.

L: As the visionary scientist Augustine said, "Who can deny that things to come are not yet? Yet already there is in the mind an expectation of things to come."

R: Whose mind? Not mine. Ah-choo!

L: God bless you.

R: Thank you. I'm getting a cold. By tomorrow I'll probably be in bed with a fever.

L: How do you know what will happen tomorrow?

R: I just told you, I'm getting a cold. I have all the symptoms—sneezing, chills. . . .

L: So the future, tomorrow's fever, exists, in a way, in the symptoms you can detect right now.

21

R: Oh, I see what you mean. But suppose someone did say that they had observed all the signs, and could tell that the earth was going to break out in an enormous feverish eruption. I wouldn't want to know about it.

L: Then you'd really be in an absurd position.

R: How so?

L: You'd be like a man holding a sieve under a he-goat while someone else tries to milk it.

R: (Laughing) Poor guy would be risking getting kicked by that goat!

L: Exactly. And all he has to do is to open his eyes to the obvious, and stop following in the path of the other man's error, to prevent it from happening.

R: It all seems less frightening when you put it that way.

L: On the contrary, it's all the more frightening.

R: Now you've got my curiosity. What is going on? How can I find out?

L: A good way to start is to listen to what science and prophecy have been able to tell us about ourselves and the world we live in.

R: Science and prophecy don't mix.

L: Wrong. . . .

Picture science and prophecy as a game of leapfrog moving through human history. Each bounds over the other, enlightened by the other's latest discovery.

Today, science can invent what in the past was entirely in the realm of prophetic vision. Take lasers, for example. The laser is both the creation and the symbol of the intuitive power of seers, that discerning power of light which enables us to reach into and describe the future. But the laser has by now become a fact of science. The light of scientific reason has, in a sense, been directed by the visionary light, has taken over and developed an area of knowledge made visible to it by prophecy.

Yet, although prophecy has, from Biblical times to the present, enlightened us about the future, we remain unenlightened about prophecy and shroud it, instead, in the darkness of our ignorance. For, although the prediction of future events fascinates us, our fascination is always tainted by skepticism and fear. Although we desire to know the events we may confront in the future, we dread the prophecies which bring us this knowledge. A passion for knowledge, as many great philosophers tell us, is always accompanied by a passion for ignorance.

Ingrained with a sense of regularity and stability, we face the future with the concern that our lives might be changed or upset by the unexpected. If we could but know what awaits us, we would be able to prepare for it, to expect it. Yet we are uneasy with the prospect of knowing what will happen tomorrow. This uneasy, eerie feeling stems from our belief that life is composed of separate, finite events, a belief that ignores the interconnection of events to those that precede and those that follow. When we turn off the light at night, we tend to relegate that day's activities to the files of our memory. They are over and done with. We forget that what we did today will have some effect tomorrow. We deny the interrelationship and continuity of life, and live in a present which is divorced from the past or the future. We should realize instead that each moment is not an isolated point, but a flooding of the moment before and after it.

Each of us has experienced occasions when we sensed what was to happen before it actually occurred. By concentrating our attention on a situation or problem, we have often surprised ourselves by correctly imagining the solution to that problem. But how did we piece together the fragments of the situation, and evoke images of the consequences? We used our minds.

The human mind is an instrument whose vast capabilities still lie beyond man's comprehension. The human mind differentiates man from the animal which is ruled solely by its instincts. The human mind affords man the ability to reason, to choose, to exercise free will—in short, it allows him to project the future consequences of any action.

Obstacles, however, do arise to restrict the ability we have to stretch, to unfold our reasoning power into the future. One

23

such obstacle is our own mortality. Because man sees the death of his physical body as an absolute termination of life, he generalizes his own finiteness into a principle of life itself. Whatever lives must die, whatever starts must finish. Thus we chop up life into units that are separated by unbridgeable gaps of death. Life, then, is thought to be lived within fixed boundaries, tightly controlled. We try to delay the end by clinging to past thoughts, to patterned and conditioned habits. We invest our energy in maintaining the status quo, in reinforcing the safety rails of the already known. Rather than experiencing reality, we cast a net of definition over it.

Men and women must, of course, continue to define, but these definitions, our understanding, can only be complete if they combine specific observations and generalized concepts. The definitions we use must be able to expand and contract, to embrace the new and simplify the redundant. Definitions must answer to data received, and not just extend themselves, heedless of what it is they define. If a balance is not maintained, if the inquiry proceeds in the exclusive direction of either the specific or the general, the knowledge obtained is limited. The overly specific advances little beyond previous knowledge, and the overly general becomes abstract and unanswerable to verification.

These two extremes of inquiry are characteristic of the different approaches to life situations which are assumed by the Occidental man and the Oriental man. The Oriental man, of Eastern culture, has offered mankind the perspective of faith, the revelation of natural laws through intuitive insights, the spiritual aspect of man's being. Yet, the Eastern man's overemphasis upon the illusory quality of matter and the perfect unity of the world suffers from a certain inability to cope with the reality of the physical world, to take responsibility for his material development and growth. The Occidental man, of Western culture, on the other hand, has offered mankind the perspective of the reasoning mind and its practical application in order to understand the physical world through investigation, analysis, and discrimination. But whereas Western scientists have made remarkable strides in understanding physical matter and using that knowledge for their own benefit, their advances suffer from their inability to bring the dissected, analyzed parts back into the whole, to completion, to unity.

24

Today we are beginning to find a burgeoning respect for the complementary mode of inquiry. Eastern philosophy is now respected by Western scientists for its understanding of the deep, fundamental principles which inform matter. Western technology is now appreciated by Eastern mystics, who see in scientific discoveries the continual confirmation of their philosophies. Although in Western society, especially, learned responses and educational systems have led to overspecialization of knowledge, truly significant discoveries have resulted more often from the complementary use of the objective and subjective, the analytical and the intuitive. The most remarkable discoveries, discoveries like those of Einstein's, have been made when a scientific model, burdened by centuries of concrete confirmations, yields finally to a new, more adequate model which is an intuitive visualization of all that the older model kept hidden. Newton's gravitational model, for example, supported by long years of evidence, was superseded by Einstein's model of relativity, which made visible all the contradictions and unexplained phenomena for which the older model had not accounted.

While some critics continue to deny intuition a place in scientific research, Fritjof Capra, a contemporary researcher in theoretical high-energy physics, insists on its importance in *The Tao of Physics:*

> "Rational knowledge and rational activities certainly constitute the major part of scientific research, but are not all there is to it. The rational part of research would, in fact, be useless if it were not complemented by the intuition that gives scientists new insights and makes them creative."

One of the best-known legends of scientific discovery through the use of intuitive, imaginative association is that of Archimedes. Asked to determine whether a crown was made of pure gold or contained a silver alloy, Archimedes was stymied by the problem. As he sat in his bath, he observed the water, displaced by his body, overflow the tub. "Eureka!" he cried, for he had indeed found it, the answer which came to be known after him: Archimedes' Principle. What he observed was that a solid body immersed in a liquid is buoyed up by a force equal to the weight of the liquid displaced by the body. A floating body displaces its own weight in a liquid. He had found the answer to his immediate problem. As gold and silver are of different densities, and, therefore, would displace different weights of water, Archimedes could test the crown for its purity of gold content. But more importantly for the long-term effects, Archimedes had found a natural principle which radically advanced man's understanding of physics.

The nineteenth-century German chemist, Friedrich Kekulé von Stradonitz, was writing a chemistry textbook when he conceived of the benzene ring, a theory which revolutionized organic chemistry. Kekulé himself has written of his process of discovery:

> "I turned the chair to the fireplace and sank into a half sleep. The atoms flitted before my eyes. Long rows, variously, more closely, united; all in movement wriggling and turning like snakes. And see, what was that? One of the snakes seized its own tail and the image whirled scornfully before my eyes. As though from a flash of lightning, I awoke: I occupied the rest of the night in working out the consequences of the hypothesis."

Once again, imaginative association provided a breakthrough in the development of a theory which dramatically advanced man's knowledge. In acknowledgment of the importance of the subjective faculty to his discovery, Kekulé ends the story of his theory's conception by urging other scientists: "Let us learn to dream, gentlemen."

But whereas in the above examples the intuitive discoveries were quickly confirmed by the practical applications and scholarly research which followed them, many other intuitions do not meet so readily with acceptance or confirmation. When this happens, the person who makes the discovery is treated like a madman and his ideas are ignored as ravings. Nikola Tesla was one such figure. The abstract scientific theories which he proposed at the turn of the century were so imaginative, so far beyond the capabilities of the technology of the time, that many of them were dismissed as pure fancy, in the West, at least. In the Soviet Union, on the other hand, his ideas were met with serious consideration and extensive scientific investigation.

One of the theories which Tesla proposed was the principle of telegeodynamics, that is, the theory that the earth could act as a conductor of mechanical vibrations. Tesla believed that with the aid of a "magnifying transmitter," he could generate high-frequency vibrations and transmit them great distances through the earth. He could thereby achieve large-scale effects on areas and objects which were situated at a tremendous distance from the transmitter. It is believed by many scientists that the Soviet Union's investigation of this theory has allowed them to develop a directed beam, very much like Tesla's original magnifying transmitter, that can be used for such diverse activities as the triggering of earthquake activity along stressful areas, creating an electrical atmosphere which is disturbing to the human mind, and developing a weapon which dematerializes physical objects.

It is not our purpose here to test these allegations. Rather, we prefer to begin the more important task of awakening the reader to possibilities which are normally excluded by the rigidity of everyday thought. The future exists outside of the everyday, and, therefore, it should come as no surprise that any knowledge we receive of it in the present will appear to us extraordinary. We must learn to examine the dreams of visionaries in order to see what they can tell us of what remains hidden from us in our waking life.

In the next chapter we shall thus examine one of the best-documented cases of prophetic vision by one dreamer, Edgar Cayce, who was known as "America's sleeping prophet."

27

II

THE PROPHECIES OF CHANGE

During the first half of the twentieth century, Cayce used the gift of trance clairvoyance to diagnose the ailments of the sick and to record insights into the mechanics of reality. Though a man of such little education that the sophisticated would class him as illiterate, Cayce, while "asleep," or in a trance, demonstrated an omniscient understanding beyond the insights of the educated around him. Originally, his readings dealt with medical questions. While in a trance, Cayce found he could diagnose the ailments of others, and that he had an uncanny ability to describe accurately a person's physical condition, often when medical doctors had given a contrary diagnosis. Although Cayce had no knowledge of medicine whatsoever, his readings were precise in their anatomical and physiological terms. These cases are fully described and may be found in the bibliography of this volume.

Although Cayce's medical predictions were enormously successful and important, not only for the specific cures which they brought about, but also for the hope they offered patients who had been abandoned by the known medical men—his powers extended beyond the field of medicine. He has, in addition, been credited with predicting such events as the First World War, the stock market crash of 1929, the Great Depression, the Second World War, and the racial strife of the 1960's. An interesting prophecy of Cayce's, verified by later events, was one he gave in August 1926, regarding weather conditions:

"As for the weather conditions, and the effect same will produce on various portions of the earth's sphere, and this in its relation to the conditions in man's affairs as has often been given, Jupiter and Uranus influence in the affairs of the world appear the strongest on OR ABOUT October 15th to 20th when there may be expected in the minds, the actions—not only of individuals, but in various quarters of the globe—destructive conditions as well as building. . . . Violent wind storms—two earthquakes, one occurring in California, another occurring in Japan—tidal waves following, one to the southern portion of the isles near Japan."

And what were some of the conditions experienced during the month of October, 1926? As recorded by the U.S. Weather Bureau, October, 1926, was an abnormally stormy month. On October 14 and 15, typhoon-force winds swept across the Kuril Islands in the Pacific. Between October 15 and 18, a tropical storm churned through the waters of southern Asia. And on October 20, one of the worst tropical storms to hit Cuba struck the Caribbean island. On October 22, an earthquake with three principal shocks shook California. Prior to the California quake, three relatively mild earthquakes hit Japan on October 19 and 20. Although Cayce had predicted *tsunamis* (or tidal waves) as consequent to the earthquake activity, no *tsunamis* followed. Despite an error regarding specific detail, the accuracy of Cayce's prediction is startling. And it is this reasonable degree of accuracy shown by his predictions that leads us to ask of Cayce and his "readings": what did he foresee for the days that lie ahead of us?

The earth changes Cayce foresaw for the future are well outlined in two of his other prophecies. In one, given on January 19, 1934, he stated:

"The earth will be broken up in many places. The early portion will see a change in the physical aspect of the west coast of America. There will be open waters appear in the northern portions of Greenland. There will be new lands seen off the Caribbean Sea, and DRY land will appear. . . . South America shall be shaken from the uppermost portion to the end, and in the Antarctic off Tierra Del Fuego, LAND, and a strait with rushing waters. . . .

The greater portion of Japan must go into the sea.

The upper portion of Europe will be changed as in the twinkling of an eye.

Land will appear off the east coast of America.

There will be upheavals in the Arctic and in the Antarctic that will make for the eruption of volcanoes in the torrid areas, and there will be the shifting then of the poles—so that where there have been those of a frigid or semi-tropical will become the more tropical, and moss and fern will grow.

And these will begin in those periods in '58 to '98. . . .''

In another, given on August 13, 1941, he stated:

"As to conditions in the geography of the world, of the country, changes here are gradually coming about.

Many portions of the east coast will be disturbed, as well as many portions of the west coast, as well as the central portion of the United States.

In the next few years, lands will appear in the Atlantic as well as in the Pacific. And what is the coast line now of many a land will be the bed of the ocean. Even many of the battlefields of the present (1941) will be ocean, will be the seas, the bays, the lands over which the new order will carry on their trade one with another.

Portions of the now east coast of New York, or New York City itself, will in the main disappear. This will be another generation, though, here; while the southern portions of Carolina, Georgia, these will disappear. This will be much sooner.

The waters of the lakes (Great Lakes) will empty into the gulf (Gulf of Mexico), rather than the waterway over which such discussions have been recently made (St. Lawrence Seaway). It would be well if the waterway were prepared, but not for that purpose for which it is at present being considered.

Then the area where the entity is now located (Virginia Beach) will be among the safety lands—as will be portions of what is now Ohio, Indiana and Illinois and much of the southern portion of Canada, and the eastern portion of Canada; while the western land, much of that is to be disturbed in this land, as, of course, much in other lands.''

Cayce believed that certain land areas would be especially danger-prone during the changes to the earth. In the United States, he felt Los Angeles and San Francisco, both of which lie along the notorious San Andreas fault line, to be prime candidates for total devastation. After the destruction along the west coast, New York would be susceptible along with the coastal regions of Connecticut and New England. The Carolinas and Georgia would be inundated, and the Great Lakes would empty into the Gulf of Mexico. Old coastlines would fall away, and there would be new coastlines further inland. Cayce also mentioned safe areas in North America, including Norfolk and Virginia Beach, Virginia; parts of Ohio, Indiana and Illinois; and most of the southern and eastern parts of Canada.

Other areas liable to volcanic and earthquake activity would include the so-called Ring of Fire in the Orient. Japan was declared prone to suffer severe destruction. The South Pacific was subject to seismic activity, as was South America, which Cayce forecast would be shaken throughout. The upper portion of northern Europe would be under water. There would be oceans and bays where there was now land. Land would appear in the Pacific and in the Caribbean off the east coast of America. There would be a change in the seasons, of frigid climates becoming tropical, and of the onset of another ice age.

Cayce never set an exact date for his forecasted events. Instead, he stressed the role of men and women in determining the events of the future. Although he was always fairly accurate in his predictions, he declared repeatedly that he did not believe in predestination. He protested that by his gift he could only see the various possibilities of the future; it was the ac-

31

NEW LAND AREAS

INUNDATED AREAS

AREAS SUBJECT TO
SEISMIC ACTIVITY

Map of the world showing areas due for earth changes, according to Edgar Cayce's predictions.

tions of men and women that must intervene to fix these possibilities at a point. Cayce described the interdependent ordering of the microcosmic world of human emotions and the macrocosmic world of the universe in his predictions:

"How do anger, jealousy, hate, animosity affect thee? Much as that confusion which is caused upon the earth by that which appears as a sunspot. The disruption of communications of all natures between men is what? Remember the story, the allegory if ye choose to call it such, of the Tower of Babel.

... For, as ye do it unto the least, ye do it unto the Maker—even as to the sun which reflects those turmoils that arise with thee, even as the earthquakes, even as wars and hates, even as the influences in thy life day by day. Then, what are the sunspots? A natural consequence of that turmoil which the sons of God in the earth reflect upon same."

All that may seem as unrealistic or fanciful in this reading will soon be dissipated in subsequent chapters in which we will consider the importance of sunspots and their possible effects on earthquakes, volcanoes, and weather patterns. We will examine not only the importance of sunspots on meteorological conditions, but also the evidence that suggests that sunspot activity exerts an influence on the psychical processes of men and women.

Edgar Cayce considered the period in which we now live, the second half of the twentieth century, a period of transition between the Piscean age and the Aquarian age. During this period he reported that we could experience forces whose natures are vastly conflicting. We would be beseiged by undiagnosable diseases of the mind and body. Adversity would test our knowledge and strength to their limits. At the same time, opportunities for tremendous scientific and technological advances would present themselves to us. We would, in brief, be at last capable of the greatest successes and the greatest failures in our dealings with other individuals, nations, and natural forces.

Cayce never saw any of the changes which he predicted as isolated, but rather as coordinated by an entire schema of change. The earth changes, he said, are caused by "movement within the earth and the cosmic activity of other planetary forces and stars.'' The movement within the earth which he cites as a primary cause is an amazing recognition of a force which was unknown to the scientists of his time. Since then, this force has been detected and studied and given the name "continental drift.'' Later we will explain in detail how this continuous movement builds up pressures and tensions which demand our careful and immediate attention.

It is almost uncanny not only how accurate Cayce's premonitory description of our age has turned out to be, but also how remarkably similar was the premonition of Johann Friede, a thirteenth-century Austrian monk. As is quoted in the *Fate of Nations,* by Arthur Prieditis, Friede forewarned:

"When the great time will come, in which mankind will face its last, hard trial, it will be foreshadowed by striking changes in nature. The alternation between cold and heat will become more intensive, storms will have more catastrophic effects, earthquakes will destroy greater regions and the seas will overflow many lowlands. Not all of it will be the result of natural causes, but mankind will penetrate into the bowels of the earth and will reach into the clouds, gambling with its own existence. Before the powers of destruction will succeed in their design, the universe will be thrown into disorder, and the age of iron will plunge into nothingness."

When we come to deal with weather changes and climate modification, we shall note how for 20 years man has used techniques to affect the weather. With assistance from man's inadvertent actions, the atmosphere may well become mixed with poisonous particles. These chemicals will eventually poison the land and air. As these chemicals make their way through the ecosystem, man will ingest them through his foods and will poison his own body system. Climate changes and a polluted environment will reduce the yields of crops and the nourishment in our foods. As in biblical times, widespread famines and epidemics of unknown diseases will plague the world population.

As we will see later, science agrees with prophetic vision that adding to mankind's misery in these forthcoming days are

34

drastic geological changes predicted for our earth. Earthquakes and volcanoes will devastate the earth. Continents will be torn apart, lands submerged beneath the oceans, and new lands rise from the oceans depths. The frequency and destructiveness of these earthquakes, volcanoes, and the consequent *tsunamis* will gather force as we approach the end of the century. The erupting volcanoes will discharge poisonous gases into the atmosphere. The sun's light will be blocked from the earth's view. Those living things that have survived these disasters will be devitalized. Indeed, terror may take hold of mankind, with many preferring death to the horror of earth life.

Many skeptics will say that we, as living individuals, are no more in a period of crisis than we, as members of the entire history of human existence, have ever been. That a whole tradition of apocalyptic literature has predicted the End since the Beginning. This time is no different from all others. In response, we remind these skeptics wherein this difference lies. Apocalyptic literature has always set the End as a destination outside of time—a place at which the present will arrive despite itself. This book is written not as such a prophecy of doom, not to terrify the reader into frustration or despair, but to place the end once more *inside* time, within the reach of the powers and knowledge of men and women. The end which we will speak of is a scientific conclusion to irresponsible human acts, and not a mythical place. It is a conclusion which is within our control and will come about or be put off by our efforts. For it is our own efforts, our own reckless technologies which make ours truly different from all past ages, which has made the end a possibility in the first place.

It was a physician Alkmeon who noted, while speaking with Aristotle, that men die because of their inability to join the Beginning and the End. This wisdom is still valid today. If we could only make ourselves aware of the connection which our past and present activities have with our future, we could attain the period of prosperity and peace for which we have been blindly struggling for so long. If we could scientifically, rationally understand our present situation, our future would be illuminated and assured. It is with this hope in mind that we will attempt in the following chapters to explore the scientific theories and facts and relate them to the earth changes ahead.

III

ASTROLOGY AFFIRMS

R: Did you ever hear of Franz Mesmer? I've just been reading about him. He used to cure people by hypnotizing
them with his eyes. Crowds of sick people came from all
over to be cured, and he would walk among them wearing
a coat of lilac silk and carrying an iron wand. What a
charlatan he must have been.

L: Yes, I've heard of him, but I wouldn't call him a charlatan. In fact, he was a very well-respected man in Parisian
medical circles during the eighteenth century. Benjamin
Franklin also paid a lot of attention to his ideas.

R: What ideas?

L: The ideas he wrote in his book, *The Influence of the Planets on the Cure of Diseases*. He believed that the planets
exerted an influence on living matter through the medium
of a universal fluid which has quasi-magnetic powers.
Every living individual was supposed to contain this
fluid—he called it animal magnetic fluid. When the fluid
was magnetically balanced, people were healthy; when it
was not, they were sick.

R: Astrological influence, magnetic fluids: this still sounds like a hoax to me. I can't believe he wasn't thrown out of the medical circles you say respected him.

L: Eventually he was. But you shouldn't be so hasty in your rejection of him. The scientists of that time were not. It took them a very long time and much investigation to finally declare, in 1784, that animal magnetic fluid did not exist. Up until then, not only Mesmer, but some of the most advanced doctors and scientists believed that it did.

R: What happened then? Without the existence of magnetic fluid, what happened to all those cures?

L: The cures continued. Medicine, hypnosis, astrology still obtain results even when the explanation of them is unclear. It is often the case that results are obtained first, and lead then to the explanations which come much later.

R: What you're saying seems to make sense, but what about Mesmer's involvement with astrology?

L: Mesmer may have been one of the most colorful exponents of astrology, but he was not its inventor. Astrology has a history which extends long before and after him. It appears, in fact, to predate recorded history.

R: Tell me more....

Various writers trace the beginnings of astrological science to the Chaldeans and Sumerians, but virtually every ancient civilization indicates an understanding of the correlations between the planetary positions and their effects on earthly affairs. In order to survive, ancient man needed a great awareness of his surroundings. A vital link existed between past civilizations and natural forces—a link which if broken would mean death. For, unlike modern man, with his conveniences of temperature control within his shelter, of food provision through market exchange, of interdependency on his fellow beings, our ancestors were more self-reliant. Through observation, man began to recognize the relationships between the heavenly bodies and his circumstances. He devised architectural constructions to observe the passages of the celestial bodies and to provide a channel between himself and the heavens. We see this illustrated in the ziggurats of Sumeria, Glastonbury, the legendary tower of Babel, Stonehenge, the temples of South America, and in the Great Pyramid at Gizeh.

Based on the law of correspondence, astrology recognizes the unity of our solar system and the interdependency of its component parts. The ancient adage, "As above, so below; as below, so above," indicates the causal relationship between the macrocosm (the universe) and the microcosm (the individual). While the Bible expresses this correspondence between the heavens above and the earth below by describing man as the image of God, ancient man reversed this reflection by anthropomorphizing the heavenly forces, that is to say, by making their gods in their own image. We see such a process of mutual reflection at work in the descriptions of astrological science. Here the zodiac, a spatial plotting of the heavens, is divided up into these twelve signs: Aries, Taurus, Gemini, Cancer, Leo, Virgo, Libra, Scorpio, Sagittarius, Capricorn, Aquarius, and Pisces, which represent different functions or phases of a cycle. From the natal chart (which includes the date, time, and place of birth of an individual) the astrologer determines the corresponding planetary locations in the signs and their geometrical configurations to give a remarkably accurate description of a person's temperament and character. History and legend relate these analogies between the archetypal temperaments of man and the twelve signs of the zodiac. It is these analogous relationships which underlie the twelve tribes of Israel, the twelve apostles of Christ, the twelve knights of King Arthur's Round Table, the twelve labors of Hercules.

While astrology has served as a useful tool for thousands of years, some modern scientists scoff at a serious consideration of the subject as a result of their *opinion* that astrology is only a carry-over of the superstitions of "primitive" man. As Bart J. Bok, a critic of astrology, wrote in the September/October 1975 issue of *The Humanist* magazine:

> "Twice I suggested to my friends on the Council of the American Astronomical Society that the council issue a statement pointing out that there is no scientific foundation for astrological beliefs. Both times I was turned down, the principal argument being that it is below the dignity of

a professional society to recognize that astrological beliefs are prevalent today."

Note that Bok calls merely for a statement of repudiation and not a scientific investigation either to repudiate or confirm the hypotheses and validity of astrological science. Even though some scientists have assumed the attitude of an ostrich with its head in the sand, hoping that if they ignored it, astrology would go away, others, researching areas outside of astrology *per se,* have concluded that there is indeed a relation between planetary influences and earth conditions. Isaac Newton, one of the founders of our modern scientific view of the universe, was knowledgeable in the astrological sciences. When Halley, "discoverer" of Halley's Comet and a doubter of astrology, questioned Newton on his deep interest and belief in the subject, Newton retorted, "Sir, I have studied the subject, you have not."

Perhaps the most apparent physical effect of extraterrestrial influence upon our earth is that of the sun and moon on tidal action of the oceans. In 1962, a correlation between the moon's phases and the intensity of rainfall was discovered by two groups of scientists, each working independently of the other. From their studies of data from 1500 weather stations between 1900 and 1949, they found the heaviest rains occurred most often in the weeks between the new moon and the full moon. The moon's influence is not, however, limited to ocean tides. From his records of tonsillectomy cases, Dr. Edson Andrews of Tallahassee, Florida, found there was a greater possibility of hemorrhaging at the full moon. Similarly, during this time fire departments report an increase in fires, police departments an increase in violent crimes, and hospitals an increase in caseloads (especially as a result of freak accidents.) Studies of birth frequencies indicate most births occur at the new moon and the full moon, while the fewest births occur a day or two before the first and last quarter of the moon.

Why should the moon's phases play such an important role in human circumstance? One possibility lies in the high liquidity of the human body, which enables the moon to exert an influence on it similar to its effect on ocean tides. Concerned with the biomagnetic fields of living beings, Dr. Leonard J. Ravitz of Duke University tested the influence of the moon upon humans. From his observations since the 1950's, Ravitz has gathered evidence that fluctuations in electrical output by human beings coincided with lunar phases, and he witnessed extremes of mood swings occurring at the new and full moon.

An impressive example of the relation between earth conditions and extraterrestrial influences is found in the work of John H. Nelson, an electrical engineer hired by RCA to study the possibility of a connection between disturbance in radio transmission and celestial phenomena. While it was known that sunspots interfered with radio transmission, it seemed that there might be additional causes for the interference and more information could be obtained from the heliocentric positions of the planets, the relationship of the planets to the sun. After checking past records of radio disturbance, he was able to conclude that magnetic storms created interference in radio transmission. The intensity of the storms correlated with specific angles formed by the relative position of the planets to the sun. As he reported in the March 1951 issue of *RCA Review*, most magnetic storms resulted when two or more planets were in 0°, 90°, or 180° relationships to the sun. He also found that disturbance-free transmission was more likely when planets made 60° or 120° angles in their solar relationship. The more planets involved in such geometrical configurations, the greater the disturbance or lack of disturbance, respectively. What is of special interest to us in Nelson's findings is the support given to astrological science, which considers the 60° angle (known as a sextile aspect) and the 120° angle (known as a trine aspect) to be "harmonious" aspects. According to astrology, the relationship of two energies, symbolized by the planets, in a harmonious aspect allows for a free flow of those energies, since there is little resistance to their expression. The 90° angle (known as a square aspect) and the 180° angle (known as an opposition aspect) however, are considered by astrologers to be "inharmonious," in which, either because of friction or a crystallized antipathy, the energies are set in direct opposition to one another. The 0° angle (known as a conjunction aspect) is a neutral aspect and can be either beneficial or detrimental, depending upon the planetary energies involved in that conjunction. Nelson's findings to this extent parallel and support the teachings of traditional astrology, which were first recorded in the *Tetrabiblos* of Claudius Ptolemy in the second century

40

A.D. Nelson's accuracy in predicting "radio weather" is gauged at between 80 to 93%, and his work has been employed in charting storm paths in advance of NASA (National Aeronautics and Space Administration) space launches.

Another study of planetary configurations and their possible effect on earth conditions was carried out by Dr. Rudolph Tomaschek, a geophysicist at the University of Munich, who once served as chairman of the International Geophysical Society. Tomaschek analyzed 134 earthquakes with magnitudes above 7.7 on the Richter scale. Tomaschek, in an article published in 1959, in the British science journal *Nature*, concluded that the inharmonious configurations of the planets, particularly Uranus, Pluto, and Jupiter, often coincided with major quakes. As we shall see in detail in the pages ahead, two scientists, John R. Gribbin and Stephen H. Plagemann, after furthering these investigations, have warned of the possibility of devastating earthquakes for the period between 1982 and 1984. At that time, the planets of our solar system will be moving into alignment on one side of the sun.

Sunspots have for a long time been known to fluctuate in 11-year cycles that coincide with the orbits of Jupiter. When Jupiter is perihelion (the point at which its orbit is closest to the sun), sunspot activity is at a maximum. But recent discoveries indicate intensification of sunspots is affected by the placement and angular configurations of the other planets as well. It is therefore believed that the concerted alignment of the planets between 1982 and 1984 will cause a maximal exertion on the sun's tidal forces and will intensify sunspot activity. The intensification resulting from this extraordinary planetary condition will occur at the same time as the normally expected peaking of the sunspot cycle. Consequently, sunspot activity is likely to be extreme during this period. According to Gribbin and Plagemann's forecasted scenario, gigantic magnetic storms on the sun are likely to result. These gigantic storms and solar flares will affect the earth's upper atmosphere, and, in turn, will disrupt global weather patterns. Since atmospheric circulation exerts a frictional effect on the earth, the disruption of global weather patterns may alter the earth's rate of spin, which, in turn, could affect stresses and strains within the earth itself— perhaps triggering earthquakes throughout the world.

After the First World War, A. L. Tchijevsky, a Russian

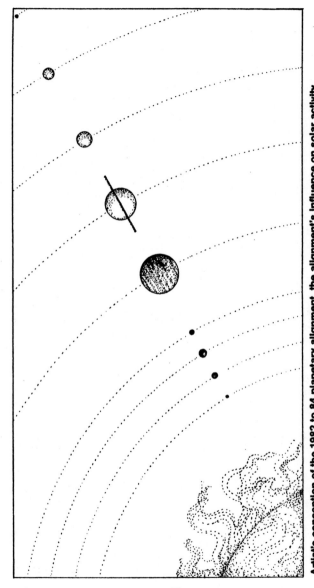

Artist's conception of the 1982 to 84 planetary alignment, the alignment's influence on solar activity, and its consequent effects on the planets are indicated.

historian, produced an Index of Mass Human Excitability. From his studies of the statistics and histories of 72 countries between the years 500 B.C. and 1922 A.D., Tchijevsky concluded that mass excitement fluctuated in 11-year cycles that he found corresponded to sunspot cycles. Mass excitement was highest at times when there was maximal sunspot activity. In his book *Earth, Radio And The Stars*, published in 1934, Harlan True Stetson, former director of the Cosmic Terrestrial Research Laboratory at the Massachusetts Institute of Technology, reported similar findings from his research into sunspots. It was Stetson's firm belief that the sunspot cycle affected the collective mood of mankind. In 1936, Harvard astronomer Loring B. Andrews announced an apparent parallel between sunspot activity and economic problems, international crises, and wars.

The studies to date do not prove for all these claims a definitive relationship between sunspot activity and earth phenomena. We should not take this lack of definitive proof as indicative of a lack of relation between the sun and the earth. This relation is not at all in question. Rather we must recognize that the isolation of a one-to-one correspondence between a specific cause and a specific effect is a dream of simplicity which science has long ago abandoned. Causes act in concert with one another, and effects are the results of a multitude of determining influences.

From the beginning, there has always been another cycle which astrology has described and on which it has based its projections. Variously called the Great Year, the Sidereal Year, or the Platonic Year, this 25,920-year cycle charts the phenomenon known as the precession of the equinoxes. Due to gravitational pulls from the moon and the sun on our ellipsoid-shaped earth, our earth tends to wobble as it revolves around the sun. If one were to extend a line from the earth's axis outward, that axis would describe a circle in space. But from the earth, it looks as if the heavenly bodies are revolving through the heavens, while the earth remains motionless. If we were to watch the relative movement of the sun and the constellations year after year, it would appear that the sun is moving at a slower rate than the constellations. Recognize, however, that we are considering apparent movement, movement as seen from our earth perspective. Noting the position of the sun relative to a fixed star, we would find, after approximately

25,920 years, the sun and fixed star to be in the same relative position they occupied 25,920 years earlier. Every 72 years it takes the sun one day longer to reach the same relative position as the fixed star.

According to astrologers, the Platonic Year is divided into twelve phases which correspond to the twelve signs of the zodiac. Every 2160 years the sun retrogrades through one phase (or age, as it is called) which represents a particular cultural, sociological, and geophysical epoch. As we noted in the previous chapter, astrologers have described the period in which we are living as the transitional period between the Age of Pisces and the Age of Aquarius. The astrological description narrates exactly the conflicts which Cayce and Friede forecast. Once again we are warned that chaos and confusion will be rampant, while the old order gives way to the new. Organizations and institutions that have functioned in the past are liable to crumble as they become outmoded for the demands of the new era. Hard pressed to find security in a rapidly changing world, people may find their very existences threatened. Mental imbalance is likely to increase. Those unable to deal with a world in flux may commit themselves to slow suicide by escaping reality with the Piscean vices of drugs and alcohol. Man might foresake reason, and revert to the animal instinct of survival for the fittest. His conflicts will lead to further aggression and wars, possibly even a catastrophic world war. But not only will man be at odds with himself and his fellow beings—the earth, too, will be torn by strife. A cataclysm to the earth is predicted, its form dependent upon which interpreter one heeds. However, whatever the causes of this cataclysm, its effects are believed to include changes in the earth's crust and consequently earthquakes, volcanoes, and *tsunamis*. The earth's climate will change drastically, altering weather patterns worldwide. Storms will intensify their devastations. The world, our world, will be shaken to its very foundations.

From an astrological perspective, the prospects for our immediate future are indeed dire. We may ask ourselves, if these forecasts of both man-made holocausts and natural cataclysms come true, will humanity continue to exist? Most astrologers believe a new form of society will be built from a nucleus of peoples willing to divorce themselves from the standards, methods, and goals of our present culture in order to realize the potential of the forthcoming Aquarian Age. Due in large measure to the adversities foreseen for the impending future, the competitive urge will of necessity give way to an increasing sense of cooperation. The community spirit will thrive, and man will finally look upon his fellow man as a brother.

The Aquarian Age is viewed as a golden age, a period in which man can make dramatic discoveries in scientific knowledge and understanding natural law. Already, we witness the remarkable technological advances of the twentieth century. These achievements are but a prelude to the anticipated revelations in the future. As man gains awareness of the underlying basis of reality, he will grow in recognition of his place within the totality of the unfolding process. Comprehending the meaning to his life's purpose, man will acquire a greater sense of security. The personal salvation he has pursued outside of himself—in religious institutions, philosophical systems, and tangible possessions—he will find to lie within himself. The Aquarian Age is termed an age of enlightenment. But to arrive at it will demand the perseverance of every individual during the trials of this present transitional phase.

Once again we are reminded that the burden of the future lies not with the seers or astrologers, who by visualizing it forewarn us of its possibilities, but with ourselves, who will realize the vision as a concrete reality. Astrology teaches us of the cyclical nature of events. A cycle can be viewed as the path which carries us back to the chaotic beginnings of a universe in which men and women had no place—or as a spiral by which we can ascend to greater harmony with the natural forces with which we coexist.

IV

THE EARTH HAS A BODY ALSO

We begin with the assumption that—kick it, build on it, bomb it, dig it up, carve it out, dam it up, let loose on it billions of industrious, careless, struggling people—the earth remains solid, unchanged, firm beneath our feet. Our everday experience confirms our belief. But science tells us that this is a fiction and that the earth is, indeed, moving under our feet.

Not so long ago, civilized people believed that the earth was flat. There was then a concept of absolute up and absolute down, and no need to think about why things fall. No need to conceptualize an invisible force such as gravity. Civilization has progressed beyond this point. We now know that the earth is round and exerts a pull called gravity on the bodies around it. But some civilized people fail to realize that invisible forces, pushes as well as pulls, stresses as well as strains, move the earth beneath their feet every bit as dramatically as psychological forces, stresses, and strains move them. In fact, there is a whole system of forces, chemical bonds and breaks, energy excitations and dissipations which bind living men and women to the living earth which surrounds them—and is not, after all, merely *under* their feet.

46

Yet anyone who has watched the changes underfoot, "changes of progress," over a twenty-, ten-, or even five-year period, has witnessed the assault of an insensitive civilization on what it mistakenly takes to be an insensitive earth. One classic example of this insensitivity can be seen in the development of various areas of Florida. Playing high-stakes "Monopoly," developers have bought up tracts of land and carved them up so as to maximize their profits. Swamp lands vital to the native ecosystem, but of little value to the real estate developer, have been covered over with fill from dredged canals. These waterways provide the developer with a checkerboard of waterfront lots, for which prospective home buyers will pay a handsome premium. Such tactics have wrought vast changes in the ecosystem. It is impossible, because of the interrelatedness of the components of an ecosystem, to change a part without changing the whole. There are no temporary or isolated changes—they are always, to a greater or lesser extent, permanent and far-ranging.

Earlier we noted that it was a synthesis of intuition and empirical knowledge which led to the scientific discoveries which have revolutionized our understanding and our relation to life. We also indicated the lack of such a synthesis which characterizes and inhibits much of our everyday thought. While ancient civilizations considered themselves a part of the land they inhabited, and watched over it as one might watch one's own health, modern civilization has detached itself from the land in order to gain a mastery over it. Today we segment the land into pieces called "property" which we can own, sell, and, in short, master in the same way that we segment our knowledge into unrelated disciplines which we also hope to master.

In studying our earth environment, scientists have tended to fall into the same pitfalls as the general public. For our scientists are products, like the general public, of societal conditioning of the educational process. Scientists have been taught to dissect and analyze, to define a limited area of knowledge as their own. Specialists have gained control in virtually every field at the expense of generalists, those who seek to synthesize specialized studies into a complete framework. While many earth scientists have recognized this problem and have pushed at their boundaries to extend their field of operation, there are still some who resist this expansion of critical attention. Such resistance is particularly detrimental to the study of ecosystems which are themselves the exact site of diverse elements, the knowledge of which can not be contained by any one discipline.

While our earth is composed of many diverse ecosystems, taken as a whole, it comprises an ecosystem within our solar system. Thus, changes among any of the components of the solar system have an effect on the earth. As we have seen, one of the most familiar effects of this sort is the tidal action of the oceans, created by the gravitational pull on the earth by the moon, the sun, and, to a slight degree, the other planets. But although most of us are aware of ocean tides, not many of us know about earth tides. According to one study reported in 1970 by a group from Columbia University's Lamont-Doherty Geological Observatory and led by John T. Kuo, the land rises and falls an average of 12 inches (or 31 centimeters) twice a day. An interesting aspect of this report is the difference in magnitude of tidal action in various regions. Measuring the earth tides at twenty different locations in North America, the Lamont-Doherty group found that earth tides on the Pacific coast were approximately 4% lower than in the continenal interior (11.5 inches or 29.21 centimeters instead of 12 inches or 30.48 centimeters), while those on the Atlantic coast were about 4% higher than in the continental interior (12.5 inches or 31.75 centimeters instead of 12 inches or 30.48 centimeters). The difference was explained by the coincidence of the earth tides with the ocean tides. On the Pacific coast, high earth tides correspond with high ocean tides. Therefore, the weight of additional water dampens the effect of the moon's pull on the land. However, on the Atlantic coast, because of the shape of the North Atlantic Basin and consequent circulation of waters, high earth tides precede high ocean tides by about eight hours. Therefore, there is not the weight of the water to restrain the effect of gravitational pull, and the coastal land on the Atlantic rises more than the average.

Some scientists believe that earth tides are influential in shaping regional geological structures. While the tidal actions in the oceans and on the land do appear at the present to have an effect on the earth's structure, it is likely that this effect was greater in past times when the moon was closer to the earth.

During these earlier times, more pronounced tidal actions probably contributed to the radical alterations of the earth which are evidenced by records of past geological periods. Later, we will be considering some of these drastic changes which occurred over the earth's 4.5 billion-year history; we are concerned in this chapter with the structure of the earth as it exists in the present.

From a surface perspective, the earth can be divided into three parts: (1) the atmosphere, (2) the hydrosphere, and (3) the lithosphere. The atmosphere is a gaseous shell that envelops the earth. Largely composed of nitrogen, oxygen, argon, and carbon dioxide, the atmosphere circulates the phases of heating and cooling between the tropics and the polar regions. Atmospheric conditions produce our weather and, in a broader sense, the climates of our earth. The atmosphere also acts as a protective shield for us by filtering a large proportion of ultra-violet radiation from the sun's rays and by burning up or slowing down the solid matter of cosmic debris, such as meteorites. Without the atmosphere, life could not exist in its present form on our planet. It should, therefore, be clear, as we will discuss in later chapters, that changes in the atmosphere will have significant effects on earth life.

The hydrosphere comprises all the water on our planet. The oceans cover approximately 71% of the earth's surface. In the southern hemisphere, the ocean waters cover roughly 81% of earth surface, while in the northern hemisphere, they cover about 61% of the surface. The earth is made up of five oceans: the Pacific, Atlantic, Indian, Arctic, and Antarctic. Sea water contains large quantities of mineral matter and dissolved gas. While it is believed that all the chemical elements can be found in the oceans, those in highest proportion include chlorine, sodium, sulfur, and magnesium. Since the temperature range in the oceans fluctuates less than on land, the oceans help to regulate the climates. By erosion, transportation, and deposition, the rivers and oceans of the hydrosphere serve as powerful agents in shaping the geological features of the earth.

The lithosphere is the outer shell of our planet, the land masses on which we live, and the ocean basins beneath the seas. As no one has yet travelled to the center of the earth, except by means of the fictional journeys of Jules Verne and other authors, our knowledge of the earth's interior is based

49

largely on analogy, geological evidence, and seismic wave studies. Because the components of our solar system are believed to have had a common origin, scientists have examined meteorites for insights into the earth's composition. The meteorites are thought to be the cosmic debris of an exploded planet, the remnants of which orbit the interstellar space between Mars and Jupiter as the belt of asteroids. Scientists have theorized an analogy between the components of the meteorites, mainly, nickel, iron, and rock known as peridotite, and the substances which make up the earth. Geology has also provided data on the earth's interior through the analysis of bore holes and of subsurface rocks brought to the surface by violent earth movement. But perhaps the greatest advance in our understanding of the earth's structure has been achieved through the study of seismic waves.

Earthquakes are a natural phenomenon of the dynamics of our planet and provide a release valve for the built-up stresses and strains within its interior. When an earthquake occurs, seismic waves are generated from its point of origin, the hypocenter of the earthquake. There are three kinds of seismic waves, primary, secondary, and long waves, known in abbreviated form as P, S, and L waves. Each of these waves has different characteristics. The P and S, known as body waves, travel through the earth's interior. By recording the time it takes them to travel to various seismograph stations around the world, we can reliably determine the earth's inner composition. The P waves are compression waves, pushing and pulling the particles they pass through forward and backward. They travel through any medium and travel at speeds of about 3.4 to 8.6 miles (or 5.47 to 13.84 kilometers) per second. The S waves are shear waves which twist the rock particles perpendicular to the direction of their movement. They travel only through solid material at speeds of about 2.2 to 4.5. miles (3.54 to 7.24 kilometers) per second. In contrast to interior waves, the L wave is a surface wave created by the S and P waves. The L waves, and their diverse forms of Rayleigh waves and Love waves, originate at the earthquake's epicenter, the point of the earth's surface directly above the quake's hypocenter. L waves travel along the surface at speeds of approximately 2 to 2.5 miles (3.2 to 4 kilometers) per second and tend to cause most of the visible earthquake damage. Through the study of seismic waves, corroborated and amplified by other evidence, scientists have proposed their conception of the earth's structure.

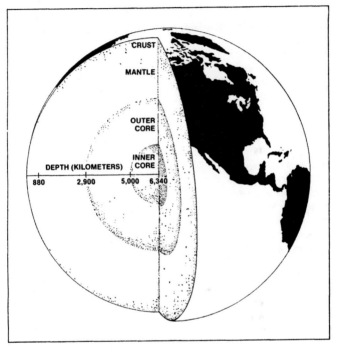

The layers of the earth.

From a structural perspective, the earth is divided into three different major components—crust, mantle, and core. The crust is the uppermost layer, a thin veneer estimated to be around 20 to 25 miles (32 to 40 kilometers) thick under the continents, and only about 3 to 5 miles (4.8 to 8 kilometers) thick under the oceans. The crust is thinnest under the oceans, considerably thicker under the continental regions, and thickest under the continental mountain ranges. Under mountain ranges, the crust is believed to be almost twice as thick as under continental regions. Analogous to the submerged portion of icebergs, mountains are believed to extend their roots far beneath the surface into the mantle. Under the oceans, the crust is composed of basaltic type rock, rocks which have been called sima due to their large concentration of silicon (si) and magnesium (ma). Under the continents this heavier type rock makes up the bottom layer of the crust. But above this is a crustal layer composed of light granitic type rock, rocks that have been termed sial because of their large concentration of silicon (si) and aluminum (al).

The crust and its differentiation from the earth's mantle was defined by the Croatian seismologist, Andrija Mohorovičić. While studying records from seismograph stations in Europe for the earthquake of October 8, 1909, in the Kulpa Valley in Croatia, Mohorovičić discovered that at depths of about 20 miles (32 kilometers) below the surface, the seismic waves increased their speed sharply. It seemed as if at this depth the waves were passing through denser rock. Mohorovičić had found the boundary where the crust ends and the mantle begins. This boundary was named after him—the Mohorovičić Discontinuity, or the Moho. The Moho is at shallow depths below the oceans basins, at greater depths under the continental regions, and at greatest depths beneath the continental mountain ranges.

The mantle extends about 1800 miles, 2897 kilometers, in depth. Our knowledge of the composition of the mantle derives from the study of seismic waves, particularly the P and S waves, which increase their speed gradually with depth. Since at different levels some body waves are reflected back to the surface scientists believe the mantle is itself layered into three sections, which they have named the lithosphere, the asthenosphere, and the mesosphere. The lithosphere includes the crust and the part of the upper mantle that has cooled and become rigid. Beneath the lithosphere is the asthenosphere, the part of the upper mantle where the rocks assume the characteristic of plastic flow. Below the asthenosphere and upper mantle is the lower mantle, also known as the mesosphere. Since both the P and the S waves travel through it, the mantle is considered a solid. Scientists believe that the mantle is composed of dense rocks of olivine (an iron-magnesium silicate), peridotite (composed of olivine and pyroxene), and eclogite (composed of garnet and pyroxene).

The core is the center of the earth. Our understanding of the earth's core is based on studies of seismic waves, analogies to meteorites, and laboratory experiments on varied conditions possibly pertaining to the earth's center. In 1914, the boundary between the mantle and core was discovered by the German geophysicist Beno Gutenberg. This boundary is called the Weichert-Gutenberg Discontinuity. As seismic waves pass from the mantle into the core, P waves reduce their speed and S waves do not pass through it at all. From this evidence the core was assumed to be molten, in a liquid state. However, in 1936 the Danish seismologist Inge Lehmann discovered that the core consisted of two different parts: a liquid outer core and a solid inner core. For while the S waves were eliminated, and the P waves slowed down upon entering the core, the P waves abruptly increased their speed at a depth of about 1350 miles (2173 kilometers) into the core, thereby indicating a transition in matter from liquid to solid. The outer core is estimated to be 1350 miles (2173 kilometers) thick and composed largely of molten iron. The inner core is believed to have a radius of about 790 miles (1271 kilometers) and to be a combination of iron and nickel alloyed with a lighter substance. Among the materials suggested as the alloy are silicon, sulfur, or sulfur and potassium.

Although such a clear delineation of the earth's structure may seem to confirm a belief in its stability and rigidity, we will consider, in the following chapter, recent concepts of the earth's structure which reveal the earth to be a dynamic, regenerative body of energy. A revolution has occurred in the earth sciences in recent years. This revolution is propelled by the theories of Continental Drift and Plate Tectonics.

V

FOREVER IN MOTION: OUR MOBILE EARTH

Would you believe the Sahara was once located at the South Pole? Can you imagine that this desert region with temperatures upward of 130° Fahrenheit (54° Celsius) was once an area with the coldest temperatures? Although such an idea may seem incredible, evidence of glacial grooves in the Sahara indicate that about 450 million years ago this desert region was once located near the South Pole. Perhaps even more phenomenal to those of us who assume that the earth's regions are permanent and stable is the theory that the continental masses are actually rafts floating across the surface of the earth. At various times in the past, continents have crashed into one another, forming mountain chains, triggering volcanoes and earthquakes, altering dramatically the earth's surface. While new land masses have arisen from the ocean depths, some continental regions may have disappeared from the face of the earth. According to the theory initially known as Continental Drift and more recently referred to as Plate Tectonics, the present surface features of the earth are the results of these processes of change. Continental Drift views change as an integral factor of the earth's dynamics in the past, present, and future. Until recently, few scientists would have hazarded an opinion for continental drift. Yet, at the turn of the twentieth century two men propounded theories of just such a phenomenon. The man most intimately connected with the initial proposal for continental drift was the German scientist, Alfred Wegener.

Wegener was not the first to consider the possibility of the earth's lack of permanency in surface appearance. Others before him had looked at the earth's surface with quizzical interest. In 1620, Francis Bacon remarked in his *Novanum Organum* on the similarity in shape of South America and Africa. But he carried the thought no further. In the mid-seventeenth century, the Frenchman François Placet suggested that the earth had been divided into the Old and the New Worlds at the time of Noah's Flood. This idea of the earth's surface having been torn apart at the time of the Flood was also suggested in 1756 by the German theologian Theodor Lilienthal, who commented on the coastline fit between the Old and New Worlds. In 1800, the German explorer Alexander von Humboldt proposed that some catastrophic action had carved out the valley which was now the Atlantic Ocean. To support his contention, von Humboldt not only drew upon the geometrical fit between the Atlantic coastlines but also on the correspondence between mountain chains in Brazil and the Congo, and between the plains of the Amazon and Guinea. With the publication in 1858 of Antonio Snider-Pellegrini's work, *La Création Et Ses Mystères Dévoilés* ("The Creation And Its Mysteries Explained"), an understanding of the similarities of the continents on either side of the Atlantic progressed from geometrical to geological fits. Citing affinities of rock formations and fossils of the Carboniferous period (about 300 million years ago) between North America and Europe, Snider believed that the two continental regions had composed one land mass in the past. He accepted the separation of the continents as having occurred at the time of the Flood.

During the 1870's and 1880's, fossil discoveries seemed to indicate that some land bridge had existed between India, Africa, and the other southern continents. The Glossopteris fern, which had flourished during the Permocarboniferous period of about 230 to 300 million years ago, was found in India, South Africa, South America, and Australia. In 1885, the Austrian geologist Eduard Suess advanced the idea of a continent formed of portions of these continents. In his four-volume work published between 1885 and 1909, *Das Antlitz der Erde* ("The Face of the Earth"), Suess called his continent Gondwanaland, after a geological province in east central India known as Gondwana, or the land of the Gonds, an aboriginal tribe. In Vol. I of his work, the geologist included southern and central Africa, Madagascar, the Indian peninsula, and Australia as part of Gondwanaland. By Vol. IV, he removed from the list Australia, linking it to another land mass. As we shall see in the following pages, the idea of Gondwanaland is now fundamental to the terminology of continental drift, despite the fact that its initial makeup has undergone extensive alterations. Suess cannot be claimed as a proponent of continental drift, for he believed that much of the land mass of Gondwanaland sank beneath the oceans from the cooling of the earth and its subsequent contraction.

On December 29, 1908, at a meeting of the Geological Society of America held in Baltimore, Maryland, Frank B. Taylor, an American geologist associated with the U.S. Geological Survey, proposed the horizontal movement of the earth's crust to explain the pattern indicating lateral compression in the mountain regions of Eurasia. In the initial stages of his thesis, Taylor gave scant consideration to the mechanics necessary for continental movement. Although Taylor's concept was revolutionary, the paper he published was not, and thus it was unsuccessful in commanding the attention of the scientific community. Independently of Taylor, Alfred Wegener, to whom we have already referred, put forth the idea of continental drift. In 1915 he formally published this idea in his book *Die Entstehurg der Kontinente und Ozeane* ("The Origin of Continents and Oceans"), which was revised through 1929. Wegener's approach was more detailed, and his views were more revolutionary, thus earning him the distinction of pioneering the theory of Continental Drift.

According to Wegener, at the end of the Carboniferous period (about 280 million years ago) there was only one continent, a supercontinent composed of all the present-day continents. Wegener called this continent Pangaea, from the Greek word meaning "all earth." Wegener believed that Pangaea started to break up during the Jurassic period (then estimated at

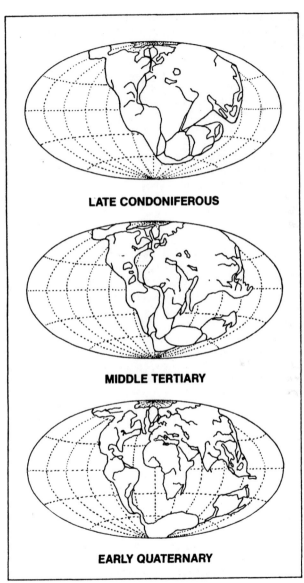

LATE CONDONIFEROUS

MIDDLE TERTIARY

EARLY QUATERNARY

Alfred Wegener's reconstruction of the progressive breakup of the continents.

40 million years ago, but now considered about 150 million years ago) and that it has continued to fragment in successive stages over the last million years. As the continental pieces separated, they moved westward and toward the equator. Wegener believed that the movements of the continental pieces created mountains by compression, leaving oceans and islands in their wake. Wegener drew upon extensive evidence from a number of different scientific disciplines. Nigel Calder characterizes him in his work called *Restless Earth*.

"Wegener was a generalist, one of those rare men who do not fear to learn and use branches of knowledge in which they are not formally trained, in order to arrive at a greater synthesis: conversely, over-specialization among Wegener's opponents blinkered their imagination."

The concept of continental drift first entered Wegener's mind after he noted the congruence of the coastlines on either side of the Atlantic. Using tracing paper, Wegener found he could make a remarkable fit between the opposing two coastlines on the Atlantic. He offered support for such a fit by geological correspondences. The two provinces of the Karoo System of South Africa and the Santa Catharina System of Brazil were found to have similar rock formations. The Cape Mountains of South Africa continued into the Sierras de la Ventana of Buenos Aires in Argentina. The Caledonian mountain system of western Europe matched the Appalachian mountain system of eastern North America.

Wegener also produced paleontological and biological evidence to support his thesis. We have already mentioned the distribution of the Glossopteris fern in the southern continents. Wegener noted the findings in South Africa and Brazil of the fossils of the small reptile Mesosaurus, known to have lived during the Permian period about 250 million years ago. For biological evidence of earlier land connections between the continents, Wegener pointed to various examples, including the following:

the manatee, found only in the tropical rivers of South America and West Africa;
marsupials, found only in Australia and the Americas;

the helix pomata, a garden snail, found in western Europe
and eastern North America;
an earthworm found in Eurasia and the eastern United
States;
perches, found in the freshwater lakes of Eurasia and
eastern North America;
the hippopotamus, found in Africa and on the island of
Madagascar;
lemurs, found in Madagascar, East Africa, Ceylon, India,
and southeast Asia.

The similarity of fauna living in diverse regions demanded some
form of land connection to allow these animals to move from
one location to another. Wegener believed continental drift
best explained the intercontinental connection.

Paleoclimatic evidence indicated radical changes in climatic zones. In the polar region of Spitsbergen, Norway, fossilized
palm trees hinted at a tropical climate in the past. Coal deposits
in Antarctica told the same story. Glaciation deposits in Africa
and South America offered a reverse shift, from a polar to a
subtropical climate. From the ice age of 280 million years ago,
patterns of glaciation in Africa, South America, Australia, and
India added support to Wegener's belief that these continents
were once all joined together. As we shall see shortly, Wegener
explained the alterations of climatic zones as evidence that
polar wandering had occurred.

Wegener also employed geodetic evidence to support
continental drift. From measurements of the longitude of
Greenland relative to Europe, Wegener inferred a westward
movement of Greenland. The records from 1870 to 1907 seemed
to indicate an appreciable separation between Greenland and
Europe. However, observations carried out after Wegener's
death found no evidence of movement, and this finding was
used by the scientific establishment to cast serious doubt on the
theory of continental drift.

Curiously enough, the greatest argument used against the
drift theory was the mechanics Wegener himself proposed for lateral movement of the continents. Although Wegener conceded
that he knew no obvious and complete solution to the question
of the mechanics for drifting, he speculated on two possible
forces: *polfluchtkraft* (pole-fleeing force) and tidal attraction.

Early in the century, the Hungarian physicist, Roland von Eötvös, suggested that the earth's spin and its ellipsoid shape causes a slight force which has the effect of moving the continental masses towards the equator. Wegener believed that paleoclimatic studies verified this pole-fleeing force. For the westward-moving force, Wegener suggested tidal attraction from the moon and the sun. By their braking action, tidal forces slow down the earth's rate of eastward rotation. Wegener felt that tidal attraction would have the greatest effect on the surface layers of the earth, thereby slowing down the continental masses relative to the interior of the earth. As a result of the differential drag, the continental masses would appear to be moving westward. Wegener had postulated the mechanics which explained for him the apparent continental movements—the *polfluchtkraft* for the equatorward drift and tidal forces for the westward drift. In the final edition of his book, Wegener suggested convection currents as a possible mechanism for continental movement.

The underlying foundation of continental drift, the possible lateral movements of continents over the earth's surface layers, was diametrically opposed to the accepted geological principle of stabilism. In 1846, James D. Dana had declared the permanency of the continents and the oceans. The geological establishment had accepted Dana's concept as fact. Wegener was from Germany, where some scientists were moving toward adopting mobilist views, but stabilism was the dominant belief in English and American geophysical thinking. Wegener's notion did more than counter the stabilist theory of continents. His theory also raised questions about another major principle in geology: the doctrine of uniformitarianism. As depicted at the end of the eighteenth century by James Hutton, the Scottish naturalist, the uniformitarianist believes that the forces operating in the present serve as an adequate explanation for changes in the past. By the twentieth century, however, the impact of catastrophic change had given way to the more gradual change of uniformitarianism. Needless to say, rigid adherence to uniformitarianism could act as a block to any "new" evidence or knowledge that might oppose accepted doctrine.

In proposing his theory of continental drift, Wegener had set a difficult task for himself. He postulated a concept that departed radically from traditional doctrine, and perhaps tried too hard to make his case. Drawing upon various scientific disciplines, Wegener attempted to be comprehensive to give credence to his theory.

Unwilling to question their own beliefs, scientists from different disciplines disputed Wegener's evidence and attempted to undermine his theory of continental drift. Wegener's geodetic evidence for the westward movement of Greenland relative to Europe was reversed by later measurements that showed no such evidence of movement. The congruence of the coastlines was called a mismatch. Geological similarities were said to be similar only from a grossly generalized perspective. Considered in intricate detail, such evidence of geological fit was considered questionable, at best. To refute Wegener's paleontological and biological evidence, the "antidrifters" explained the necessary land connections of the past by means of land bridges. According to this theory, land bridges once spanned the oceans, linking certain continents, thereby allowing the migration of species from one continent to another. After the Cretaceous period (about 60 million years ago), the land bridges sank beneath the oceans, cutting off these intercontinental routes. Similarities of flora on different continents were explained by possible migrations of seed and spores carried by air or by ocean currents.

Although the attacks against Wegener's theory precluded initial acceptance, thus respectful inquiry into, continental drift, the theory did suffer from one gross inadequacy: the mechanics he postulated for the driving force of continental movement. It was this area, the mechanics of continental drifting, that became the largest stumbling block to the theory's acceptance. On this specific issue, a formidable opponent was the English geophysicist, Harold Jeffreys of the University of Cambridge, England. From his calculations, Jeffreys showed that both forces proposed by Wegener, the *polfluchtkraft* and tidal actions, were insufficient to power the alleged continental movements. Jeffreys estimated these combined forces to have only one-millionth the power necessary to propel the continental masses. The weight of the scientific community was against Wegener's theory. Continental drift was considered a spurious concept, born of a lively imagination.

Wegener's advocacy may have seemed a lone cry in the wilderness, but there were others in the scientific community who risked their reputations by arguing the plausibility of

61

continental drift. Wegener's voice was silenced when he died during a 1930 expedition in Greenland. Continental drift was again championed by the South African geologist, Alexander L. du Toit. In his book, *Our Wandering Continents,* published in 1937 and dedicated to Wegener, du Toit expanded upon Wegener's ideas. Du Toit provided further geological evidence of intercontinental connections in the past, including what he called the Samfrau geosyncline, a Paleozoic feature stretching from Argentina to Australia to southern Africa. While du Toit accepted Wegener's underlying thesis of continental drift, he proposed two primordial continents, as opposed to Wegener's supercontinent of Pangaea. Du Toit called these two continents Gondwana and Laurasia. Composed of South America, India, Africa, Australia, and Antarctica, Gondwana was in the southern hemisphere; in the northern hemisphere was Laurasia, made up of Europe, Asia, North America, and Greenland. The "Tethys" Ocean separated the two continents until they began to break apart. Ascribing a contributory but minor influence to Wegener's *polfluchtkraft* and tidal forces, du Toit supported the view of radioactive heating within the earth's interior and consequent convection as the driving force behind continental movements.

The theory of convection within the earth's interior was proposed in 1928 by Arthur Holmes of the University of Durham, England. A pioneer in the use of radioactivity to measure geological time, Holmes believed that convection currents were formed within the earth from the decay of the radioactive elements in the rocks. As the rocks grew hotter from this radioactive decay, Holmes felt they would assume a plastic quality and rise towards the surface. Near the surface the heat would spread out laterally. When the rocks cooled, they would sink back into the earth's interior as cold, dense matter. Holmes suggested that convection currents could be the driving force behind continental movements. According to the general scheme of his theory, Holmes considered the oceanic ridges to be the products of the ascending currents of the mantle convection. The convection currents descended with cold material along the oceanic trenches. As a result of this process of ascending and descending currents, the continental masses were carried along the earth's surface as if on a conveyor belt. This concept of Holmes anticipated the later findings in the 1950's and

1960's that would substantiate "sea-floor spreading." During the same period that Holmes proposed his theory, two other scientists expressed the belief that convection was a reality of earth dynamics. In his theory of thermal cycles, John Joly, an Irish scientist, enlisted convection from radioactive decay as fundamental to mountain building. This is included in the second edition of his book entitled *The Surface History of the Earth*, which was published in 1925. In the 1930's, Fleix Vening Meinesz, the Dutch geophysicist, advocated thermal convection as a process operating within the earth's interior. He based this on the gravity anomalies he found over the ocean trenches and island arcs in the western Pacific. The theory of convection in the earth's interior met with resistance, since recent seismic evidence had indicated the earth's mantle to be solid.

While the die-hard opponents of continental drift continue to this day to focus their criticism on the mechanics proposed for drifting, the evidence accumulated from various scientific disciplines has led to general acceptance of continental drift by the scientific community. Initially, the evident congruence of the two coastlines on either side of the Atlantic had aroused interest in the plausibility of all the continental regions once being joined together. Fits of the geometrical pieces to the continental drift puzzle are still going on. S. Warren Carey of the University of Tasmania used a continental contour about halfway down the continental slopes. In 1955, he tested the fit of Africa and South America and found a convincing match. At a symposium on continental drift sponsored by the Royal Society of London in 1964, Edward Bullard, J.E. Everett, and A.G. Smith, all of the University of Cambridge, England, presented their computer study of the geographical fit of the continents on either side of the Atlantic. Using a computer to determine the best fit of the continental margins, Bullard and his assistants found a remarkable fit at a depth of about 500 fathoms (900 meters). At this depth, the margin of error in the match was no greater than one degree. Although the continental edge is thought to lie at about 1000 fathoms (1800 meters), Bullard's study showed the the best geographical fit to occur at the 500-fathom depth, where the continental slope slants steeply downward. Using a 1000-fathom isobath (a line around the continent at that depth) and including three submerged "continental fragments,'' Walter Sproll of the Environmental Science Services Administration and Robert S. Dietz of the National Oceanic and Atmospheric Administration (N.O.A.A.) found a computer match between Australia and Antarctica. Other computer studies have shown fits at depths of 1000 fathoms between Africa, Antarctica, and India. Increasing use of the computer has provided scientists with another means to test the various possibilities and parameters of continental matches.

Taking into account continental drift and its ramifications of continents once being joined together and later splitting apart, the history of fossil evidence can be readily understood. Paleontological evidence corroborates the unity of the southern continents of South America, Africa, India, Australia, and Antarctica into a single land mass (Gondwanaland) and the unity of the northern continents of North America, Europe, and Asia into Laurasia. Fossils of conifers, of which the pine tree is a member, show that different types of conifers about 70 million years ago grew either on the land masses of Gondwanaland or of Laurasia, but not on both. Freshwater animals are believed to have travelled easily between South America and Africa. Coral reefs show identical conditions for eastern North America and western Europe some 350 million years ago.

Recent paleonotological discoveries in Antarctica have helped fit that continent into the matrix of the Gondwana land mass. Although it has been known since 1900 that Antarctica contains coal deposits, and thus once had a tropical climate, it wasn't until 1967 that fossils of a vertebrate were found there. A fragment of bone was discovered in the Transarctic Mountains in December of that year by Peter J. Barrett, then a graduate student at the Ohio State Institute of Polar Studies. Edwin H. Colbert of the American Museum of Natural History in New York examined the fragment and identified it as part of the lower jaw of a labyrinthodont amphibian. Fossils of this fresh-water amphibian had also been found in South Africa and Australia. Colbert felt certain that this creature would have been unable to withstand a salt water migration. In 1969, Colbert joined a team led by David H. Elliot, also of the Ohio State Institute of Polar Studies, to search for fossil vertebrates in Antarctica.

While Barrett's find caused a sensation within the geological world, the findings of Elliot's group were even more impressive. At Coalsack Bluff in the Transarctic Mountains they found not only further evidence of the labyrinthodont amphibian but fossils of the reptilian genus, Lystrosaurus, as well. A land animal of the early Triassic period (about 230 million years ago), the Lystrosaurus had also been found in the Upper Beaufort beds of the Karoo System in southern Africa, in the Tunghungshan beds of Sinkiang, China, and in the Panchet Formation of India. The following quote is taken from Colbert:

> "The discovery of the Lystrosaurus with other mammal-like reptiles and with labyrinthodont amphibians indicated that we had found in the Transarctic Mountains an association of amphibians and reptiles similar to that occurring in the Lower Triassic beds of South Africa, designated the Lystrosaurus fauna."

This discovery provided an essential link between the fauna of Antarctica and of Africa. Since the Lystrosaurus fauna of South Africa also contained fossils of the mammal-like reptile Thrinaxodon, the finding of the Thrinaxodon in Antarctica would make the link between Africa and Antarctica even more conclusive. In 1970, James W. Kitching of Witwatersrand University of Johannesburg, John Ruben of the University of California, and Thomas Rich of Columbia University, set off for Antarctica to find the Thrinaxodon. The expedition proved successful. At McGregor Glacier, the group discovered a skeletal imprint of the Thrinaxodon. They also found fossils of other mammal-like reptiles and labyrinthodont amphibians which correspond to the Lystrosaurus fauna of South Africa.

These paleontological discoveries indicate a necessary dryland connection between Antarctica and southern Africa. A narrow land bridge between the two continents is unacceptable. Land bridges tend to act as zoological filters, increasing the diversity of fauna between the two land masses. The close resemblance between the early Triassic fauna of South Africa and Antarctica virtually demands a past connection of these two regions—integral parts of a single continental land mass.

Since the early geological finding which Wegener and du Toit used as evidence for continental drift, further findings confirm the likelihood of the two primordial continents of Gondwana and Laurasia. In matching the present-day pieces of these continents, geological evidence indicates "fits" from the topographical extension of mountain ranges and from the similarities of both the rock strata and the grain of the rock strata. While there is convincing geological evidence to link North America, Greenland, Europe, and Asia into a single land mass (Laurasia), some of the more interesting matches have focussed on the fits between the geological provinces of Africa and South America. In discussing Wegener's evidence for continental drift, we mentioned the similarity between the Karoo System of South Africa and the Santa Catharina System of Brazil and that between the Cape Mountains of South Africa and the Sierras de la Ventana in Argentina. In his work, du Toit included the Samfrau geosyncline, linking Australia to Argentina and southern Africa. Using Bullard's computerized fit of South America and Africa, Patrick Hurley of the Department of Earth Sciences, Massachusetts Institute of Technology, in 1967 announced a geological match between the geological provinces of São Luis, Brazil, and one near Accra, Ghana. Hurley's findings followed two other matches between geological provinces in Brazil and in western Africa: between Recife, Brazil, and north central Cameroon, and between Sergipe, Brazil, and Gabon. The detailed similarity of rock formations between these provinces indicates a connection of these two continents in times past. Findings from the southern continents have matched South America, Africa, India, Australia, and Antarctica into a single land mass, Gondwanaland.

Perhaps the most convincing evidence in behalf of continental drift in recent years has come from the paleomagnetic studies of continental rocks in the 1950's and of ocean sediment in the 1960's. Although man has known about magnetic rocks for 2500 years, William Gilbert, a seventeenth-century English physician and physicist, discovered that the earth itself acts like a magnet. Initially, Gilbert believed that the earth contained a giant bar magnet. He conducted high-temperature experiments with an iron bar magnet, and demonstrated that the bar loses its magnetism at high temperatures and regains it upon cooling. He concluded that the internal heat of the earth would prevent a permanent magnet from existing there. The concept of a permanent magnet inside the earth had to be abandoned. Something, however, makes the earth *act* like a bar magnet. This imaginary magnet is located at the earth's center and lines up in the same general direction of its axis of rotation, but with a slight distortion. Because the earth's magnetic axis and its geographic axis do not coincide, compass readings have to be corrected for the deviation between the two. The earth's magnetic field both surrounds and permeates the earth. Although the magnetic field is known to be in a continual state of flux, the cause of the earth's magnetic field still remains an unanswered question for geophysicists.

The most generally accepted theory for the magnetic field assumes that the earth's interior operates like a dynamo. Joseph Larmor, a mathematician and physicist at Cambridge University, was an exponent of this theory, which he suggested in 1919. The dynamo theory was fleshed out in the detailed work of two scientists working independently of each other: Edward Bullard of Cambridge University and Walter M. Elsasser of the University of California at San Diego. According to their theory of magnetohydrodynamic motions within the earth, the earth's molten iron and nickel outer core is analogous to the electrical conductors of a dynamo. Convection currents within the earth's core (estimated at something less than 0.1 centimeter per second) provide the motion, and the electric currents generated create the magnetic field. The complexity of the convection circulation patterns would also allow for the increase and decrease of magnetic force, reversals of the magnetic field, and variations in the motions. Thermal energy from the radioactive elements within the earth's core and the energy of the earth's rotation are considered to be the sources of energy for the convection currents. Some scientists believe that the relative equilibrium of the convection currents has now made the magnetic field self-perpetuating. Most geophysicists accept the Bullard-Elsasser theory. Nonetheless, the earth's magnetic field is still not completely understood.

During the nineteenth century, Pierre Curie studied high-temperature effects on various magnetic materials. His experiments were similar to Gilbert's earlier experiments of high-temperature effects on an iron bar magnet. Above a certain critical temperature, which varied according to the type of materials, Curie found that magnetic materials become non-magnetic. When they cool, these materials regain their magnetic properties with the magnetization set in the direction of the magnetic field in which they are cooled. The critical temperature point is now known as the Curie point. As a result of Curie's findings, rocks containing magnetic particles provide scientists with a record of the direction and the strength of the earth's magnetic field at the time the rocks solidified. The magnetization is frozen into the rock at its formation, and is not affected by subsequent changes in the direction of the earth's magnetic field.

In 1906, Bernard Brunhes, a French physicist, found some volanic rocks magnetized parallel to the earth's magnetic field but in the opposite direction of the earth's present field. Brunhes concluded that at some time in the past, the earth's magnetic field had reversed polarity. Scientists now recognize four major epochs of alternating polarity for the earth's magnetic field over the past 4.5 million years. During each epoch there have also been short-term reversals known as events, lasting less than 100,000 years. For the last 700,000 years, (referred to as the Brunhes Epoch), the magnetic field has had a normal (north-seeking) polarity. Prior to the Brunhes Eopch there was the Matayuma Epoch of reversed (south-seeking) polarity. Before the Matayuma Epoch there was the Gauss Epoch of normal polarity, preceded by the reverse polarity of the Gilbert Epoch. The transition time between polarity reversals is believed to occur over relatively short time spans.

The earth's magnetic field has reversed itself in times past, and evidence shows that the positions of the magnetic field are constantly changing. Such findings can be explained by the wandering of the magnetic poles, by the movement of the continental regions, or by a combination of the two. This phenomenon was initially known as polar wandering. In 1891, Seth Carlo Chandler, an amateur astronomer living in Cambridge, Massachusetts, investigated the recorded variations in latitude. Chandler believed the variations to be the result of the earth's wobbling motion, which he attributed to two cycles—an annual cycle and a cycle of 428 days, a 14-month cycle. Chandler felt that the annual cycle was the consequence of seasonal fluctuations in the oceans and in the atmosphere. The second cycle, now called the Chandler wobble, still defies complete comprehension. One explanation for the Chandler wobble is the elasticity of the earth's interior. The influence of cosmic factors is highly probable. In the ionosphere, the earth's upper atmosphere, the radiation of the sun strips electrons from the oxygen and nitrogen atoms. The positive and negative charge particles of the ionosphere (the ions and electrons) make the air an electrical conductor, and the electric currents generated in the atmosphere create magnetic fields that contribute to variations in the earth's magnetic field. Scientists believe there may be yet another type of excitation involved in the Chandler Wobble. In 1906, John Milne, an English professor of geology working in Tokyo, and the inventor of the seismograph, proposed a connection between earthquakes and the Chandler wobble, and evidence does suggest some such relationship. Breaks in the polar wandering path are known to precede earthquakes. Operating on a 12-month and a 14-month basis, the two cycles sometimes work against each other, resulting in minimal motion. Every seven years the two cycles coincide and work together. At such times (e.g. 1964, 1971, 1978) the rate of motion in the polar drift increases. There also appears to be a high incidence of large magnitude earthquakes during this time.

Although the paleomagnetic evidence was initially attributed to polar wandering, in the mid-1950's researchers of paleomagnetism found the need to include continental drift as an explanation for their findings. Based on the 1954 studies of the paleomagnetism of rocks in England, P.M.S. Blackett and his assistants at the University of London proposed a 30° clockwise rotation and a northward drift of England. A year later, S. Keith Runcorn at the University of Newcastle-Upon-Tyne interpreted paleomagnetic evidence in Europe's rocks as a consequence of polar wandering without incorporating the theory of continental drift. Runcorn soon changed his opinion: studying the paleomagnetism of rocks in North America, he found a wandering curve for the magnetic pole that did not coincide with the one found for the rocks in Europe. If the two land masses were brought together, the diversity of the polar curves would be lost. Runcorn thus stated the need of continental drift to support the paleomagnetic evidence.

As the evidence of paleomagnetism accumulated from studies of rocks on other continents, the case for continental drift strengthened. As Anthony Hallam writes in *A Revolution In The Earth Sciences:*

"It was not long before a series of results were reported from rocks collected in the southern continents. Once again, a systematic change with time through the Palaeozoic, suggestive of polar wandering, was recognizable, and the divergences of polar wandering paths for different continents could be eliminated if they were brought together as Gondwanaland.''

One significant aspect of geomagnetic field changes is the effect they have upon earth life. Fossil evidence taken from ocean sediments shows catalytic evolutionary changes to earth life-forms during the transitional periods between magnetic field reversals. Independent studies by Robert J. Uffen of Queens University, Canada, Bruce Heezen and James D. Hays, both of Columbia University, New York, suggest that genetic mutations, the appearance of new species, and the elimination of other species coincide with periods of polarity reversal. These scientists propose that the reason for such dramatic changes to earth life-forms is the weakening of the protective shield provided by the magnetic field. The magnetic field and its variations affect the atmosphere and the world's climates. For one thing, they shield the earth from potentially lethal cosmic rays. When the magnetic field diminishes during transitional periods of polarity reversal, this shield does not offer the usual protection. Cosmic rays bombard the earth with radiation that can severely affect life. However, the atmosphere also serves as a protective shield from cosmic radiation and, therefore, may filter genetically damaging rays from earth life even during periods of a weak magnetic field. Although some scientists discount the effect of the magnetic field on genetic mutations, attributing the evidence of rapid evolutionary changes at such times to the consequent changes in the world's climates, it is likely that variations in the magnetic field strength do indeed have an effect on earth life-forms even though it may be very subtle.

In his research, Hays found a susceptibility to variations in the magnetic field in such species as fruitflies, mud snails, and flatworms. For humans, this effect could be one of mental disorientation. Cosmic ray radiation is composed of high-energy frequencies, which are known to have disruptive effects on human mental balance. The magnetic field is believed to have weakened by about 50% over the past 2500 years. There are indications of an impending polarity reversal, and we should be aware of the possibility of genetic mutations to life forms in the future. We may even suffer a loss in our mental equilibrium. Some fatalists go so far as to hypothesize the elimination of man during the next field reversal.

During the last 20 years, revolutionary advances in our knowledge of the world's oceans have provided further evidence in favor of the theory of continental drift. We divide the oceans into three major zones: the continental margins, the deep ocean floor or abyssal plain, and the mid-ocean ridges. In the 1950's, Maurice Ewing, of Columbia University, discovered that all the world's mid-ocean ridge systems were connected as one vast network, forming an underwater mountain chain about 40,000 miles (64,374 kilometers) long, some of which rise 15,000 feet (4,573 meters) above the deep ocean floor. Instead of being a smoothly flowing chain of submarine mountains, this network of mid-ocean ridges is sliced into segments offset at approximately 90° angles along what are known as transverse fracture zones. The crests of the mid-ocean ridges are scarred by rift valleys (cracks in the earth's crust), volcanoes, and shallow earthquakes. Molten material welling up from the mantle below oozes through the rifts. As it cools, the material solidifies and ejects earlier sediment from the crest of the ridge. Years earlier, in discussing his theory of convection and Wegener's theory of continental drift, Arthur Holmes had proposed the oceanic ridges to be possible sources of upwelling material from the mantle. The oceanic trenches would be areas in the convection circuit where cold, dense material descended back into the mantle. Employing geothermal convection to explain his findings, Felix Vening Meinesz discovered anomalies in gravity over the oceanic trenches and island arcs of the western Pacific. Later evidence confirmed the island arcs and ocean trenches to be zones of instability in the earth's crust. Like the mid-ocean system, the ocean trenches and island arcs are punctuated by volcanoes and earthquakes.

In the early 1960's, Harry Hess of Princeton University and Robert Dietz of the National Oceanic and Atmospheric Administration (N.O.A.A.) drew together the findings from oceanic explorations and presented a radical theory about the world's oceans, which they worked out independently of each other. To explain Ewing's discovery of the worldwide system of ocean ridges, Hess and Dietz rediscovered the theory Holmes

71

had suggested some 30 years before, refined it in accord with the oceanographic evidence, and published it as the theory of Sea-Floor Spreading. If the idea of the permanence of continents and oceans was threatened by Wegener's theory of continental drift, the oceanographic evidence was devastating. Sea-floor spreading resulted from the combined discovery of a worldwide network of mid-ocean ridges, the high heat flow, and the high frequency of volcanic and earthquake activity there. Hess proposed that the mid-ocean ridges are situated above the ascending currents of the convection circuit within the mantle. Along the ridges, new crust is created from the upwelling of molten material that solidifies into ocean sediment as it drops below its Curie point and pushes earlier sediment further out from the ridge crest. If the sea floor is constantly expanding from the ascending currents of convection, there should be some area where the sea floor is being destroyed, where the convection currents descend into the mantle. The oceanic trenches, as Holmes had mentioned so long before, seemed a likely candidate.

Hess noted several factors that seemed to confirm the ocean trenches as the locations of the descending currents of the convective cell, the areas where the sea floor was continually being destroyed. Vening Meinesz's discovery of gravity anomalies above the ocean trenches in the western Pacific had an important influence on his proposal of a convection cell within the earth's interior. Hess believed that the descending current of the subcrustal convection cell held down the oceanic trenches as much as four kilometers below the deep ocean floor. Although Hess considered himself a uniformitarian, he did admit his belief in catastrophic events in the earth's history. One such "great catastrophy" for Hess was the convection cell within the earth's interior, which he believed had instigated the sea-floor spreading. The ocean floor seemed to him a conveyor belt which emerged at the mid-ocean ridges, spread out on both sides of the ridge, and descended at the ocean trenches. From his 1956 observations of the flat-topped sea mounts that rise from the deep ocean floor, which he first discovered while on tour of the Pacific during the Second World War, he obtained evidence to support this belief. He called these sea mounts "guyots." Some of the guyots were tilting toward the ocean trenches, as if they were being carried along passively toward

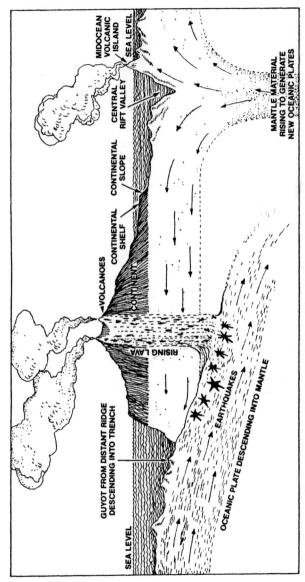

Sea-floor spreading as conceived by Dr. Harry Hess of Princeton University.

73

their eventual destruction at these trenches. He concluded that the ocean floor is always renewing itself.

To corroborate Hess's concept of sea-floor spreading, Fred Vine and Drummond Matthews, both of Princeton University, suggested that the molten material welling up along the mid-ocean ridges would be magnetized in the direction of the earth's magnetic field when it cooled below its Curie point and solidified. According to the theory of sea-floor spreading, new sediment pushes older sediment further away from the ridge. The age of the ocean-floor sediment increases as one moves away from the ridge. Since the earth's magnetic field is known to have reversed its polarity in the past, one would expect the two sides of a ridge system to have bands of sediment with alternating polarity. Ocean sediment, therefore, provides a magnetic record to date polarity and reversals, and conversely to date the ocean floors. Analyzing sediment taken from the Indian Ocean, Vine and Matthews explained the magnetic patterns they found in the ocean sediment sample and the symmetry of the bands on either side of the ridge as indications of sea-floor spreading from both sides of the ridge at an equivalent rate. Later evidence gathered from sediment taken from the Reykjanes Ridge in the mid-Atlantic, and other ocean ridges, have verified the findings of Vine and Matthews: ocean sediment takes on the magnetization of the earth's magnetic field at the time of its solidification and thereby records the history of normal and reversed polarity. It also allows for local variations in the sea floor spreading rate. These findings match the magnetic field timetable established through the study of paleomagnetism in continental rocks. The paleomagnetism of ocean sediment and its verification of sea-floor spreading provided convincing evidence to many scientists who had earlier scoffed at the idea of the movement of continents and oceans. Under the weight of accumulated evidence, the established geological doctrine of permanence was fast giving way in favor of the doctrine of earth mobility. We have discussed the convection theory of Arthur Holmes, who saw in his theory a possible force for Wegener's continental drift. We have considered the Bullard-Elsasser theory of the magnetohydrodynamics within the liquid outer core of the earth interior. Although it is an area whose total comprehension lies presently beyond man's knowledge, the earth's magnetic field is thought to be generated and maintained by convection currents in the liquid outer core. The theory of sea-floor spreading and the impressiveness of its evidence gave greater credence to the mobilist perspective of geophysics. Operating through some form of convective cycle, many of life's phenomena go through the processes of creation on the ascending currents and of destruction on the descending currents, but with the promise of continual renewal. We see convective cycles repeatedly in daily life. We know that convection currents in the atmosphere can unleash severe weather patterns, such as hail and violent thunderstorms. An obvious example of convection on the mundane level is boiling water in a pot. As the water at the bottom of the pot heats under the flame, that water rises and displaces the cooler, denser water near the surface toward the sides of the pot where it descends and, following the pattern of the convection circuit, is forced under the flame to heat and rise again to the surface. A similar process takes place in the earth's core.

Temperature increases as one descends toward the earth's core. This results from the thermal convection generated by the radioactive decay of rock minerals. Estimates of the temperature at the center of the earth range from 2000° Fahrenheit (1093° Celsius) to 6500° Fahrenheit (3593° Celsius). As shown by the high heat flow above the mid-ocean ridges, and the low heat level above the ocean trenches, convection circuits operate beneath the earth's crust, creating new ocean floor at the ridges and destroying old ocean floor at the trenches. The ocean floor operates like a conveyor belt, passively carrying everything above it toward the ocean trenches where the floor descends back into the incinerator of the mantle. Its molten material flows beneath the crust toward the ocean ridges.

The Canadian geophysicist J. Tuzo Wilson carried the concepts of continental drift, convection, and sea-floor spreading one step further. In a 1965 article published in *Nature* magazine, Wilson introduced the concept of the transform fault and proposed the idea that the earth's surface is broken into several "plates." The theory known as Plate Tectonics was born.

The worldwide system of oceanic ridges, first discovered by Ewing, is a network now characterized by segments offset from the ridge chain at approximate 90° angles. These displacement areas are known as transverse fracture zones, but were named transform faults by Wilson, due to the fact that the

75

displacement either stops or changes direction along these faults. According to Wilson's view of earth dynamics, our planet's surface is broken into several major "plates" and a number of smaller "plates." These plates contain both the continental masses and the ocean floor regions. The earth's upper layers are now divided into the asthenosphere and the lithosphere, the latter constituting the earth's crust and part of the upper mantle that has cooled and become rigid. Lithospheric plates ride along the plastic flow of the asthenosphere, the part of the upper mantle where rocks assume a plastic quality. Plates can behave in three different ways. In accord with sea-floor spreading, plates can either be created at the mid-ocean ridges or be destroyed at the ocean trenches. Wilson proposed a third movement, in which two plates slide past each other, causing horizontal tearing or shearing. Transform faults have been related to the structural geology of New Zealand, Turkey, Japan, and California, along the notorious and potentially devastating San Andreas fault.

Since Wilson's initial proposal, proponents of plate tectonics have defined a range of six to nine major plates and about a dozen smaller plates estimated to be about 100 kilometers thick. Since all the plates are interconnected, activity at one point in this mosaic of the earth's surface can have effects thousands of miles away. One possible example of this effect is the earthquake that destroyed the Chinese city of Tang Shan in July, 1976, killing approximately 750,000 people. In 1977, Peter Molnar of the Massachusetts Institute of Technology and Paul Tapponnier of the University of Montpellier, France, proposed that the continuing collision between the Indian subcontinent and Eurasia may have triggered the Tang Shan earthquake some 2500 kilometers to the northeast. Most of the activity, however seems to occur at the edges of the plates. Transform faults are now known for their earthquake proclivities.

Lithospheric plates cannot overlap. But when two plates meet, one may eventually override the other, creating an oceanic trench in which the lower plate will descend into the earth's asthenosphere. If both plates have oceanic margins, one will override the other, which could result in an arc of volcanic islands on the margin of the overriding plate. A plate with a continental margin will override a plate with an oceanic margin, due to the buoyancy of sialic material of the continental

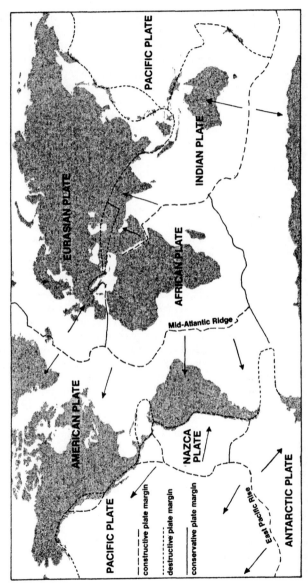

Map showing the major lithospheric plates.

PACIFIC PLATE

EURASIAN PLATE

INDIAN PLATE

AFRICAN PLATE

Mid-Atlantic Ridge

AMERICAN PLATE

PACIFIC PLATE

NAZCA PLATE

East Pacific Rise

ANTARCTIC PLATE

constructive plate margin
destructive plate margin
conservative plate margin

margin. In this situation, the force of the two plates colliding tends to compress the continental margin, creating a range of folded mountains such as those on the west of South America. Two plates with continental margins can crash into each other, but their buoyancy may prevent either one from overriding the other, and the two continental margins may join together to form an even larger continent. Such a situation appears to be occurring between the Eurasian plate and the Indian plate, with the suture being the ever-heightening Himalayas. The major features of the earth's surface are explained by the theory of plate tectonics, but the concept is relatively young and is being constantly refined by more recent studies and research. Two men significant in refining Wilson's initial concept of plate tectonics are Dan McKenzie of Cambridge University and W. Jason Morgan of Princeton University.

Wilson believed that frictional heat was generated by the melting of an oceanic margin plate overridden by a continental margin plate and that this produced permanent pockets of hot lava within the mantle. The study of permanent hot spots is generally associated with Morgan's work and is known as the Plume Theory. According to Morgan, there are some 20 stationary plumes beneath the moving lithosphere. Estimated to be some 150 kilometers in diameter, these plumes are thought to originate from convection in the lower mantle, possibly at the mantle-core boundary, and to rise toward the asthenosphere at about two meters per year. This hot molten material spreads out horizontally at the athenosphere and contributes to the movement of the lithospheric plates. The zones beneath the lithospheric plates where this material rises are punctuated by high crustal heat, volcanoes, earthquakes, hot springs, steam jets, and the like. Iceland and the Hawaiian Islands are considered products of hot spots. When the Pacific plate moved across the stationary hot spots in the Pacific area, they punched holes in the plate above, creating the volcanic islands of Hawaii. Hot spots have also been found in the ocean floor of the Red Sea. Not all the hot spots of the world are located beneath ocean floors. In continental Europe, there is a hot spot known as the Eifel Plume with its volcanic chain extending from the Eifel Mountains in Germany across Czechoslovakia to Upper Silesia in Poland. The volcanoes of West Africa are said to be caused by hot spots.

An interesting sidelight to the plume theory is the proposal put forward by Peter Vogt of the University of Southern California in 1972. A significant contributor to the study of plume activity, Vogt believes plume discharge goes through cyclical action, reaching a peak of discharge every 50 to 60 million years. It has been approximately 60 million years since the last peak of plume discharge, and plume discharge is presently on the increase. Vogt feels another peak of discharge may occur in the not too distant future. Perhaps the most important consideration of peak plume discharge for us is the effect it could have upon the climatic and faunal conditions of our environment. Volcanic activity is associated with an increase in plume discharge. When volcanic dust particles are thrown into the upper atmosphere, they tend to cut off the solar heat reaching the earth. Increases in plume discharge might also accelerate the movement of the lithospheric plates, producing more collisions between continents, thereby increasing mountain building, which consequently alters climate zones and causes greater extremes in temperatures. The destruction wrought by volcanic activity and climate changes poses a radical threat to the survival of life-forms. Allan Cox of Stanford University has suggested that changes in plume activity might also influence the magnetohydrodynamics of the earth's liquid outer core, thus affecting the magnetic field variations and the frequency of polarity reversals. We have already noted the coincidence of genetic mutations, the appearance of new species, and the elimination of other species during periods of polarity reversal. Continental drift and plate tectonics, once considered wild and irresponsible theories, are now generally accepted within the scientific community. The stabilist who followed Dana's belief of a rigid, stable earth eventually had to cede recognition of continental movement. While commendable for its underlying recognition of cause and effect, and the cyclical flow of process, the authority of uniformitarianism carried to an extreme can prevent serious consideration of new theories or hypotheses. Similar to the stabilists, those who bow solely to the authority of uniformitarianism may rigidly isolate themselves in the past, and lose innovative curiosity. In light of recent discoveries, we can now say that continental drift does fit the prerequisites for acceptance by uniformitarianism. Sixty years ago it did not. Under the weight of overwhelming evi-

79

dence, old beliefs can be proved wrong. We must always be wary of falling prey to a narrow-minded view of life's processes. Many of our beliefs based upon present knowledge will be later repudiated by the extensions of man's learning.

The mobilism of plate tectonics is gradual and interminable, thus satisfying the foundation of uniformitarianism, but there are periods during this slowly evolving process when change accelerates at an alarming rate and magnitude—periods of catastrophe. Catastrophe theorists realize that within the slowly developing nature of life there are dramatic periods of transition that shape the peaks and troughs of the cycles in life phenomena. Transitional phases are now considered a function of the earth's interior in zones where the component materials undergo drastic, radical alterations of form and characteristic. Largely due to man's perspective of space-time and to the evidence of cataclysmic events throughout the history of the earth, the idea of catastrophic change has a long tradition in geology. The episodic revolution that drastically transforms existing conditions is forever latent during the long periods of time the earth may seem quiescent.

The earth is a vibrant energy being, influenced by the dynamics of cyclical processes. At times, the impact of these cycles on earth life may not be evident, but at other times their impact is severe and cannot be escaped. The more complicated an event, the greater its range of behavior. The more variables involved, the greater their interrelationship which can accelerate geometrically. While each variable component undergoes its own cycle of behavior, all share transitional periods with the peaks and troughs of any cyclical process. When the peaks and troughs of several integral factors meet, their influence could provoke extremes of behavior. In considering the drifting of the earth's magnetic field, we noted the coincidence of large earthquakes with the seven-year cycle of polar wandering when the annual cycle and the Chandler wobble of 14 months both peak. In the concurrent peaks or troughs of several significant cycles lies the potential for catastrophic change.

The past has witnessed drastic transformations of earth life. Great floods, dinosaurs, and ice ages are part of our earth's history, despite the remoteness of it in terms of our modern life style. Ancient legends and traditions speak of flourishing civilizations destroyed by cataclysmic events. Although certain

80

rationalists might dispute the existence of the civilizations known as Lemuria and Atlantis, our present understanding of earth dynamics no longer rules out the possibility of the disappearance of continental land masses. Through continental drift, sea-floor spreading, and plate tectonics, we recognize certain processes during which continents break apart and collide to form new land masses. The sea floor is continually renewing itself. Lands rise out of and sink beneath the ocean depths.

From a larger perspective, catastrophic incidents are but slight blips on the wide screen of the total process. What might seem catastrophic to us are merely momentary releases in a gradually evolving continuum. Learning about the dynamics of the earth's structure can give us a clearer understanding and knowledge of our territorial limits within this planet. The earth reproduces itself like the creatures that come out of it. We are all part of this immense and awesome whole. Our planet is not a demented space ship forever travelling without purpose in circles around the sun, but a body in constant motion which has taken care of itself and thus far adapted to major changes throughout the eons. Should the earth continue to survive the traumas of nature's and mankind's behavior, it will look radically different a thousand years from now. It's time to turn our attention to those slight "blips" whose castastrophic actions can kill hundreds of thousands of people within moments. The following chapter deals with the effects wrought by volcanoes and *tsunamis*.

VI

THE EARTH'S FIERY VOICE: VOLCANOES

"Now in this island of Atlantis there was a great and wonderful empire which had rule over the whole island and several others, and over parts of the continent, and, furthermore, the men of Atlantis had subjected the parts of Libya within the columns of Heracles as far as Egypt, and of Europe as far as Tyrrhenia. . . . But afterward there occurred violent earthquakes and floods, and in a single day and night of misfortune all your warlike men in a body sank into the earth, and the island of Atlantis in like manner disappeared in the depths of the sea."

——Plato, *Timaeus*, 25 (as translated by Benjamin Jowett)

Throughout history, man has struggled to survive against the natural earth spasms of volcanoes, earthquakes, and *tsunamis*. The cataclysmic effects of natural disasters are recorded in all world cultures, for all world cultures have been touched by the havoc and destruction wrought during the earth's spasmodic releases of accumulated internal strains. Fascinating accounts of radical changes to the earth are told in the legends of lost continents. One such lost continent is known by the name of Atlantis.

Around the sixth century B.C., Plato recorded the legend of Atlantis in his dialogues of *Timaeus* and *Critias*. Plato describes the island empire of Atlantis, but ends his account abruptly at the point where Atlantis, which had thrived by adhering to moral virtue and natural law, falls from a state of idyllic grace into decadence through boundless ambition and perversion. Plato dates the destruction of Atlantis at about 9600 B.C.

Occult tradition, in which Plato was well-versed, associates the fall of the Atlantean civilization to the Biblical fall of man, where man abandons his spiritual perception of essential self and dons the garb of corporeal sensations and material concerns. This analogy is also viewed in the fall of the fool in the esoteric system of the tarot. Man falls from his perfect state of essential being into the material world both to experience and conquer physical temptation, being given his faculty of reason and his ability to exercise free will through alternative choice. By using free will, man can regain his spiritual elevation. The return toward spiritual development is wrought with the concerns of the material world.

The cataclysms that ended the earlier Lemurian civilization are said to have created the continent of Atlantis in the mid-Atlantic. The Atlanteans had spiritually evolved to an exalted level of conscious awareness of natural laws and the harnessing of natural forces. Their technology was highly developed. Solar power was harnessed and concentrated through crystal. Yet within the advanced state of their technological knowledge lay the seeds of their undoing. The Atlanteans focused more and more upon the material aspects of their lives, thereby losing through neglect the spiritual perception that had guided their growth. Greed and the misuse of knowledge became dominant as some members of the community sought to gain advantage first over their fellow citizens and then over other countries. Various cultures tell of visits to their lands in times past by men who "came out of the sea." (The similarities between the structures of the pyramids in Egypt and Latin America remain an awesome mystery to this very day. According to occult teachings, the leading scientists of Atlantis became aware of the possibility of catastrophic disaster to their island and emigrated to the four corners of the world.) At the height of their material power, the Atlantean civilization was snuffed out by violent cataclysms that rocked the continent and sank it to the ocean floor.

Is the story of Atlantis a fairy tale? Perhaps. Yet such an account must have been considered far more fantastic in the context of the permanency of continents and oceans some 20 to 30 years ago, before overwhelming evidence established the theory of plate tectonics with its continual creation, destruction, and renewal of the earth's surface. Within the concept of plate tectonics, the plausibility of a lost continent in days long past can no longer be denied. The process of mobile lithospheric plates allows for the lack of surviving physical evidence. The submerged continent could have eventually travelled to an ocean trench to be consumed in the mantle's fiery inferno. Although the legends of Atlantis and lost civilizations have mystified mankind throughout past centuries, today's rationalists believe they have found the physical location and the answer to the Atlantean myth. These scientists point to the explosion of the Santorini volcano 60 miles (100 kilometers) north of Crete in the Aegean Sea sometime between 1500 and 1400 B.C.

The explosion of Santorini is hard to imagine. Recent evidence indicates that it was the largest volcanic explosion known to man, an explosion estimated at twenty times the force of the Krakatoa explosion in 1883. The Santorini explosion is said to have hurled some six cubic miles (24 cubic kilometers) of ash into the atmosphere. On parts of Thera, an island remnant of the Santorini explosion, there are deposits of ash up to 100 feet (30 meters) deep. About 70 cubic kilometers of volcanic material were thrown out by the explosion. The island shattered and collapsed into the sea, forming a caldera (circular cavity) about 33 square miles (85 square kilometers) in area. Rimming this caldera were three crescent-shaped islands—the remains of Santorini. Since then, mild eruptions from the caldera center have created two other islands in the Santorini group. In 1967, a city given the name Akrotiri was discovered buried under volcanic ash on the island of Thera. Excavations suggest that the town's population, numbering approximately 30,000, had evacuated before the eruption, possibly warned of the impending disaster by preceding earthquakes.

The discovery of Akrotiri and recent evidence confirming

the magnitude of the Santorini explosion support the theory proposed in 1939 by Greek archeologist Spyridon Marinatos, who suggested that the Santorini explosion destroyed the Minoan civilization and caused a shift of power from Minoan Crete to mainland Greece. The intensity of the Santorini explosion would have unleashed *tsunamis* that may have wiped out the Minoan coastal towns. The volcanic ash would have destroyed crops and cattle, eliminating the Minoan food source. The poisonous gases might have forced emigration. The Minoan culture, a highly aesthetic and powerful civilization, may have lost the battle for survival against one of earth's natural spasms, the volcano.

The reason for volcanism in the Aegean is attributed to plate tectonics and the overriding of the African plate by the lithospheric plate on which the Aegean region moves in a southeasterly direction. As one plate overrides another, volcanic and earthquake activity is generated through the friction and intense heat. It was this process that produced the exposion of Santorini in the period between 1500 and 1400 B.C. It is this process of plate tectonics that accounts for most volcanic activity.

There are over 500 active volcanoes in the world. Volcanoes are related to the movement of the lithospheric plates, and more specifically to the releases of tensions and strains within the earth's crust. An estimated 80% of volcanoes occur along the subduction zones where one plate overrides another. Approximately 16% of known volcanoes occur along rift systems, such as the mid-ocean ridge system, discussed in the last chapter, and the East African Rift Valley. The few remaining areas of volcanicity are associated with the collision of two continental margin plates (as in the case of the Mediterranean region) and with hot spots punching holes through lithospheric plates (as in the creation of the Hawaiian Islands). Volcanoes along ocean trenches tend to be more explosive than those along the rift systems, which tend to be more effusive in their lava flow. Earthquakes often accompany volcanoes, and areas of high volcanicity are also regions of high earthquake incidence. Most volcanoes lie along one of two great global belts. About two-thirds of them occur along a belt circling the Pacific Ocean. Known as the Ring of Fire, this Pacific belt includes the volcanoes of South and Central America, the Aleutian Islands, Japan, and the Philippines. The other major belt, known as the Mesogean or Himalaya-Alpine belt, runs in an east-west direction and reaches from southeastern Europe and the Mediterranean through southern Asia to Indonesia.

Composed of liquid lava, gases, and fragmented solid material known as pyroclasts, volcanic eruptions range between the two extremes of effusive and explosive eruptions. Occurring along the mid-ocean ridges and on deep-ocean islands, effusive volcanism is the molten material of the mantle welling up and overflowing along the rifts in the earth's crust. Explosive eruptions, on the other hand, result from the rapid expansion of the gases in the magma. Volcanic material is ejected as pyroclastic matter consisting of the fragmented, solid material of pumice, ash cinder, and dust. Explosive volcanism tends to occur along the continental side of the ocean trenches where one lithospheric plate is forced under another. Whereas the volcanic rock produced at the ocean ridges is of a basalt type, the explosive volcanic rock tends to be of a kind that falls between a basalt and granite type. These rocks, known as andesites, are composed of material from the mantle, from the melting of the descending oceanic margin plate, and from the bottom of the overriding continental margin plate. Although a volcano may historically have an effusive nature, that does not prevent it from exhibiting explosive characteristics in the future. Most volcanoes combine effusion and explosion, but there are too many individual characteristics of each volcano to allow for simple generalizations. An exception may be found to any rule in the study of volcanology. While volanoes are commonly divided into the two types of volcanic eruption, effusive and explosive, they are also classified into four general categories: 1. Peléan, 2. Vulcanian, 3. Strombolian, and 4. Hawaiian.

The Peléan eruption is the most explosive type of volcano, taking its name from Mont Pelée on the island of Martinique in the West Indies. As the pressure of gases in the gas-rich magma increases, the solidified lava that plugged the vents of the volcano's crater is blown off, often horizontally, taking along with it parts of the volcanic mountain. Pyroclastic matter is ejected, along with the explosively expanding gases (composed of water vapor, carbon dioxide, hydrogen sulfide, and chlorine). While the explosive force of a Peléan eruption is frightening, there is usually no great lava flow. The most terrifying aspect of the Peléan eruption is its glowing cloud (*nuée ardente*). A mixture of gas, vapor, and dust, these glowing clouds charge down the volcanic mountain with hurricane-force speeds upwards of 100 miles (160 kilometers) an hour and extend as far as 60 miles (97 kilometers). With temperatures ranging from 800° to 1000° Celsius, these *nuées ardentes* level everything in their path, as in the tragic incident of St. Pierre.

On the morning of May 8, 1902, four explosions burst from Mont Pelée, shooting forth a glowing cloud of gas and ash which wiped out everything in its path. There were warning signs: on April 2, the first signs of activity in Pelée were noticed—the steam vents (fumaroles). On April 23, minor earth tremors were felt, and ash rained down on St. Pierre. On April 25, there were more earth tremors. The rain of ash and poisonous gases began to kill the birds. As the inhabitants of the city fled the city, people fled into the city from the surrounding countryside. The government refused to let anyone leave. There was an important election scheduled for May 10. On May 7, as the eruptions continued, the governor of the island attempted to allay the fears of the population. A government commission declared that Mont Pelée was a passive volcano. When this false campaign started to fail, the national guard was called in to stop the evacuation attempts of its terrified citizens. By May 10, there were no citizens left alive to vote. Thirty thousand of them lay buried in the wake of the Mont Pelée eruption. Only two people survived.

A Vulcanian eruption is similar to a Peléan eruption in its initial phases. The congealed vent plugs are blown out with the ejection of pyroclastic matter, and often with destruction to the volcanic structure. Although a suffocating cloud of gas and ash rises above the volcano's crater, the *nuée ardente* of the Peléan type, with its violent hurricane force, is absent. The Vulcanian eruption is likely to have more lava flow, and tends to go through cycles of explosive force followed by quieter, effusive flow through the vents. The lava in the crater eventually forms a crust and traps the expanding gases until, under increasing pressure, the gases again explode through the crust.

The eruption of a Strombolian-type volcano, like the Vulcanian type, is a blend of both explosive and effusive volcanism. In both types, the pent-up pressure of gases in the magma explodes the volcano's plugs and ejects pyroclastic matter. Both also emit lava flows. There is an important distinction between them, however. Unlike other volcanoes that go through cycles of activity and quiescence, the Strombolian eruption is constantly in action. It takes its name from the Stromboli volcano which rises 3000 feet (915 meters) above sea level and close to 7000 feet (2135 meters) above the ocean floor north of Sicily in the Aeolian Islands. Known as the "lighthouse of the Mediterranean," Stromboli has been active for over 2000 years. A basaltic volcano, Stromboli generally goes through relatively mild eruptions, with only two exceptions: in 1930 and in 1971, Stromboli experienced explosive eruptions.

The Hawaiian type of volcano has the quietest eruption, but the most effusive lava flow. The Hawaiian type of eruption is basaltic, characterized by intermittent activity when lava overflows a crater. Hawaiian-type volcanoes are known as shield volcanoes because their gently sloping broad dome structures are built up by successive lava flows. Unlike other types of volcanoes, the low viscosity of the lava in the Hawaiian volcanoes allows the expanding gases to escape with ease, and consequently with a minimum of explosive force. These lava flows are referred to as "rivers of fire," since they are known to reach velocities between 10 and 25 miles an hour (16 to 40 kilometers). A series of earthquakes usually precedes the volcanic eruption. The Hawaiian-type eruption occurs primarily in the string of volcanic islands for which it is named. It is believed that these islands were created by a permanent hot spot that punched holes through the Pacific plate moving over it.

Originating about 18 million years ago from the volcanic activity now dormant in the northwest, the Hawaiian Islands stretch southeastward to the island of Hawaii, where eruptions began about six million years ago and continue even today. Both Mauna Loa and Kilauea on the "Big Island" of Hawaii are active. Mauna Loa rises 13,680 feet (4171 meters) above sea level, and some 16,000 feet (4878 meters) from ocean bottom to sea level, constituting a total height of about 30,000 feet (9146 meters). The volcano has a lava dome 60 miles long (97 kilometers) and 30 miles wide (48 kilometers). Mauna Loa seems to erupt with cyclical regularity. According to records kept for over a hundred years, it tends to be active approximately 6.2% of the time and averages one eruption every 3.6

years. A typical eruption begins with a gas-ash cloud and lava fountains. Low viscosity lava spurts out from the vents and rifts. Great masses of lava flow for days, travelling at high speeds. The volcano eventually quiets down and falls into repose. A similar process takes place with Kilauea, located on the southeastern slope of Mauna Loa. With a lava dome 50 miles (80 kilometers) long and 14 miles (23 kilometers) wide, Kilauea rises 4090 feet (1247 meters) above sea level (almost 10,000 feet or 3,049 meters below the summit of Mauna Loa) and about 20,000 feet (6098 meters) above the ocean bottom.

Icelandic eruptions are often classified with the Hawaiian type. Unlike the Hawaiian Islands, however, the Icelandic volcanic structures are plateaus that rarely attain any great height, the tallest being about 3300 feet (1006 meters). Some measure only 330 feet (100 meters). While the major part of lava flow in a Hawaiian eruption goes to the building up of its steep dome, the lava flow in an Icelandic eruption spreads over great distances. Apart from this difference, the eruptions of Hawaii and those of Iceland are similar. The lava is basaltic and thus erupts in an effusive flow without explosive violence. The whole of Iceland is composed of volcanic material. It has been calculated that of all the volcanic material discharged in the world since 1500, approximately one-third of it has been discharged in Iceland. Sometimes, as happened in 1963, it creates new land formations, islands rising from the ocean depths.

On November 14, 1963, a submarine eruption began south of Iceland, 10 kilometers (six miles) southwest of Westmanna Ejar, the Vestmanna Islands. Through three vents along a fissure system, steam jets shot pyroclastic material 60 meters (200 feet) into the air. By November 16, a new island began to emerge from these effusions. Eruptions continued for months, and the island grew in size. This steaming new land, named Surtsey, was visited by seagulls two weeks after its birth. By the summer of 1964, insects, flies, and butterflies frequented it. This is one instance of life starting to take hold in the wake of volcanic action. The majority of volcanoes, however, leave only destruction.

One of the most devastating volcanic eruptions occurred in Iceland in 1783. Preceded by a week of local earthquakes, Skaptär Jokull (Mount Skaptar) became active in June 1783, and continued to erupt into 1784. Commonly referred to as the

89

Laki eruption, it is considered to be Iceland's worst catastrophe. On June 8, the eruption began with violent explosions and tremendous lava fountains. Volcanic ash rained over a wide area, and an effusive flow of basaltic lava covered some 580 square kilometers (220 square miles) of countryside. Farms were overrun and livestock lost. The majority of deaths were attributed not to the direct effects of the Laki eruption, but to its indirect effects, which triggered what is now known as the "Great Haze Famine." A bluish haze, produced by the volcanic ejection of an estimated 130 million tons of sulphur dioxide, hung over Iceland and was observed throughout most of the world. Referring to the smoke issuing from Iceland, Benjamin Franklin, then serving as United States ambassador in Paris, wrote:

> "During several of the summer months of the year 1783, when the effects of the Sun's rays to heat the Earth should have been the greatest, there existed a constant fog over all Europe and a great part of North America. This fog was of a permanent nature; it was dry and the rays of the sun seemed to have little effect towards dissipating it, as they easily do a moist fog. . . . Of course, their summer effect in heating the Earth was exceedingly diminished.
> Hence, the surface was early frozen.
> Hence, the first snows on it remained unmelted. . . .
> Hence, perhaps, the winter of 1783-4 was more severe than any that happened for many years."

Because of the far-flung volcanic ash of that eruption, crops in Scotland and Norway were severely damaged. In Iceland itself the effects were disastrous. Farmlands were destroyed by lava flow and volcanic ash. The remaining grass was stunted by the volcanic gases. Three-fourths of all the sheep and horses and one-half of all the cattle died. Because the haze reduced visibility, Icelandic fishermen could not work the sea, thus closing off another vital source of survival. Due to the indirect effects of the Laki eruption, close to one-fifth of Iceland's human population of about 50,000 perished from starvation and disease during the 1783 Great Haze Famine.

Perhaps the best known of all volcanic eruptions was the 1883 explosion of Krakatoa. Located in the Sunda Strait between Java and Sumatra, Krakatoa in the 1880's was an uninhabited volcanic island, whose history was largely unknown. Although there were accounts of an eruption in 1680, Krakatoa was thought to be dormant. In the late 1870's, the region of the Sunda Strait, a seismically active zone, was shaken by a series of earthquakes. No connection between the earthquakes and Krakatoa was made until May 20, 1883, when Krakatoa reawakened from its long sleep. A series of explosions were heard over 150 kilometers (90 miles) away. A steam column and a shower of ash issued forth from a vent in the volcano. Activity continued until May 27, when it quieted down. On June 19, however, Krakatoa resumed its volcanic activity. By the end of June a second column of steam and ash was seen. Throughout July explosive blasts and minor earthquakes rocked the surrounding areas of Java and Sumatra. On August 11, a third column of steam and ash was observed. Krakatoa's activity continued to intensify. On the afternoon of August 26 explosions occurred with fearful regularity until the morning of August 27, when all hell broke loose. With four violent explosions, Krakatoa erupted. The third blast, shortly after 10:00 A.M., was the most intense, and sent a cloud of ash and dust skyward to heights estimated at 80 kilometers (50 miles). Approximately 20 cubic kilometers (five cubic miles) of volcanic material were hurled into the air. The rain of pumice and ash fell over an area of some 780,000 square kilometers (300,000 square miles) with accounts of ash falling three days after eruption on decks of ships at distances of 2800 kilometers (1600 miles) away. The rain of ash darkened the skies 450 kilometers (275 miles) away from the site, and areas 85 kilometers (50 miles) away suffered continual darkness for over two days. The explosions were heard as far away as 5000 kilometers—reports of the sounds were logged at Alice Springs in central Australia, some 3500 kilometers (2200 miles) southeast of Krakatoa, and at Rodriguez Island in the Indian Ocean about 4800 kilometers (2950 miles) to the southwest. By August 28, the eruption of Krakatoa had ended. Two-thirds of the island had disappeared in the explosions. Krakatoa had collapsed in upon itself, forming a caldera.

Like a large rock thrown into a pond, with consequent waves rippling out, the collapse of Krakatoa sent waves hurtling outward from the devastated island. These waves, known as *tsunamis,* ravaged the surrounding coastlines of Java and Sumatra. Reaching heights of as much as 40 meters (130 feet), they were responsible for the destruction of 295 towns and villages and for the deaths of over 36,000 people. While the serious damage inflicted by the *tsunamis* was confined to the shorelines along the Sunda Strait, traces of these *tsunamis* were recorded as far away as the Bay of Biscay off the French and Spanish coasts, some 17,255 kilometers (10,350 miles) away.

The most significant effect of the explosion of Krakatoa, apart from the tragic deaths of over 36,000 people, was its influence upon the world's climates. The volcanic ash ejected into the upper atmosphere by the explosion formed a dust cloud and drifted westward. On the afternoon of August 27, the cloud was seen in Ceylon. The following day, it had reached South Africa, and two days later it was over the Atlantic. By September, the cloud haze had arrived over the west coast of South America, and was headed toward the south of Hawaii. On September 9, it had circled the earth and had reached its point of origin. However, the cloud did not stop there. It continued westward and drifted around the earth at least three times. While its initial circling was largely confined to a zone 15° on either side of the equator, the cloud spread out and covered most of the earth, except perhaps for the polar regions. Because of the increased particle matter in the atmosphere, the world experienced brilliant red sunsets for years. This increased particle matter also had adverse effects. At the observatory in Montpellier in the south of France, measurements over several years after the eruption showed a 10% drop in the average intensity of solar radiation. The volcanic dust blown into the atmosphere by the Krakatoa explosion caused a two-year drop in world temperatures.

As with the Laki eruption of 1783, the effects of volcanic eruptions on the world's weather was again demonstrated by the Krakatoa eruption. In both cases volcanic "dust veils" proved significant in lowering world temperatures. Because most of the incoming solar radiation is of wavelengths either equal to or smaller than the volcanic dust particles, the particles absorb, disperse, and reflect some of it back to interstellar space. The situation is aggravated by the longer wavelengths of the outgoing radiation that is reflected from the surface of the earth. Because volcanic dust particles cannot absorb or intercept outgoing radiation, there is a steady level of outgoing radiation continuing while incoming radiation is reduced. This process results in the lower air temperature which dramatically alters the world's weather. Based upon a mean value of direct solar radiation for the years 1883 to 1938, scientists have found that following volcanic eruptions in the years 1883, 1888, 1902, and 1912, the monthly average of direct solar radiation dropped as much as 20 to 22% below the solar radiation mean. Volcanic dust veils can maintain a profound influence upon incoming solar radiation for three years after an eruption. Some of the particulate matter of dust veils may stay in the stratosphere for as long as 12 years after an eruption.

Although their injection of solid matter into the atmosphere might seem unimportant to some readers, it is estimated that a fluctuation of just 1% in the incoming solar radiation can alter earth surface temperatures by as much as 1.5° Celsius. The meteorologist Hubert H. Lamb, Director of the Climatic Research Unit at the University of East Anglia, England, studied the recorded volcanic eruptions between 1500 A.D. and the 1960's, and concluded that major volcanic eruptions can have temporary but significant effects upon world climates. Other scientists go so far as to suggest a correlation between volcanic activity and the earth's glacial ages. Maintained over a long period of time, the volcanic dust veil could severely disrupt the climatic equilibrium and lead to a spreading of glaciation away from the poles. A correlation between volcanic activity and glaciation was proposed in 1974 by James P. Kennett and Robert C. Thunell, both of the University of Rhode Island. Drawing from their analyses of ocean sediment cores taken from the world's oceans, they concluded that increased explosive volcanism during the Quaternary triggered the succeeding Great Ice Age of the Pleistocene Period (about one million years ago).

While volcanic activity can trigger climate changes, some scientists assert that climate changes serve as catalysts for volcanic activity. For example, if cooling world temperatures were to induce the spread of glaciation, a great volume of the world's water would be frozen into the encroaching ice. As the ocean level falls, due to this freezing action, the equilibrium of the earth's surface features could be disturbed. Such disturbance

93

might instigate radical shifts within the earth's interior and trigger off earthquakes and volcanic activity. The concept of climate changes as the cause of volcanic activity is hotly debated within the scientific community. Such a process may seem quite farfetched for some, but there is little doubt about the far-reaching effects of volcanic activity. One of the best examples of the global consequences of a volcanic eruption can be seen in the eruption of Tambora in 1815.

On the island of Sumbawa, 330 kilometers (200 miles) east of Bali, the volcano of Tambora erupted in 1815. Although little is known about this explosion, Tambora is believed to have released about 840 times the energy of the Krakatoa eruption in 1883. An estimated 100 cubic kilometers (25 cubic miles) of volcanic material was ejected. Death estimates range up to 90,000–100,000 people killed by the rain of ash and pumice, and possibly another 80,000 dead from famine and disease on Sumbawa and the neighboring island of Lombok. The direct effects of the Tambora eruption were nothing short of catastrophic. According to the economic historian John D. Post, in his book *The Last Great Subsistence Crisis In The Western World*, which was published in 1977, the planetary crop damage and famine which resulted from the Tambora eruption triggered an extremely volatile political atmosphere.

The year following the Tambora eruption, 1816, came to be known as the "year without a summer." At weather stations in western Europe, the months of June through September showed a dramatic decrease in the mean temperatures compared to preceding years (a range of 4 to 8° Fahrenheit). In New England the average temperature for the summer was 4.4° Fahrenheit below normal. The irregular weather caused a drastic reduction in crop harvests. One report tells of an area in Wales with only three or four rainless days between May and October. Poor harvests resulted, and, with them, food shortages. Food shortages led to skyrocketing food prices. As Post records, grain prices virtually doubled. Using unweighted averages of a wholesale grain price index, with the year 1815 as a base year with an average of 100, Post indicates the price changes during the years following the Tambora eruption. Taking the total effect for western Europe and the United States, the price index went from a level of 100 in 1815 to:

140	in	1816
198	in	1817
125	in	1818
88	in	1819
72	in	1820

Although grain prices peaked in 1817, the harvest failures of 1816 could not be stopped from working their effects. Between January 1815 and June 1817, cereal prices in Bavaria jumped fourfold. In Switzerland, bread increased four or five times its 1814 price level.

To help us imagine the consequences of the harvest failures of 1816, Post calculates that agriculture accounted for between one-third and one-half the national income. One can readily understand the dire impact of harvest failures upon a nation's labor force and upon the national economy itself. The ripple effects of harvest failures intensify the tragedy. Because of heavy rains in 1816, more firewood was used; the demand rapidly outstripped the supply. Prices increased sharply in the German region of the Rhineland by November 1816. The phenomenal increases in food and fuel costs led to a dramatic slackening in the demand for industrial goods. Unemployment levels soared in the industrial sector as well as in the agricultural sector. Whole nations suffered severe declines in national income.

Famine and starvation spread rapidly through the western world. Legions of beggars wandered across Europe. The year 1817 became known as the "year of the beggars." Famines and irregular weather patterns resulted in typhus, which swept across Europe between 1816 and 1819. Starvation was the main cause of death in Switzerland. The fear of starvation had dramatic social and political consequences. Suicides increased. Mass emigrations occurred, such as the movement of Germans from their homeland to America and to Russia. Even in the United States there was movement from New England to the Midwest. In other European countries, demonstrations, looting, and rioting were widespread. Post claims that the subsistence crisis became the rallying point for the various criticisms against established authority. To counter the unrest, European governments became conservative and repressive. The effects of Tambora on world climate might have been temporary, as reflected in the return to good harvests by 1819, but its eruption seems to have had a profound and lasting influence on the course of human history.

The preceding examples of volcanic eruptions have illustrated the effects of such cataclysms on a local and a worldwide scale. Nonetheless, volcanoes are fundamental to the earth dynamics of mobile lithospheric plates. They will not disappear. In fact, if Peter Vogt is correct in believing that the present increase in plume discharge is leading to a significant peak in the future, then volcanoes may soon become even more active. Vogt suggests that plume discharge and its volcanic eruptions go through cycles of intense activity, followed by phases of relative quiet. During intense activity, a period Vogt believes we are now entering, volcanic activity catalyzes the radical transformation of evolution—transitions with mutations in and extinctions of animal life. As we have noted, the earth is a delicately balanced entity. Change one aspect, and you change the entire environment. Our earth is highly sensitive, reacting to the variable factors that influence it. With a broader understanding of the individual processes that account for the total result of the planet earth, we can better prepare ourselves for the likely earth changes ahead.

Perhaps we can't prevent volcanoes from erupting. And the unique characteristics of volcanic eruptions defy generalizations. We, however, can recognize hints and warnings of incipient volcanic activity. Most significant increases in volcanic activity are preceded by a swarm of earthquakes or ground tremors. The ground surface surrounding the volcanic structure begins to well up. Before eruption, the volcanic structure usually swells. The use of tiltmeters to record changes in the slope of the ground signals potential volcanic activity. Before eruption, the upwelling of hot magma also affects the local magnetic field. As noted in the preceding chapter's discussion of the continual creation and spreading of the ocean floor, magma ejected from the earth's mantle will take on the direction of the magnetic field in which it cools below its Curie point. Monitoring the magnetic field around volcanoes forewarns of changes in the magmatic flow beneath the surface. We can also monitor the heat flow from volcanic areas. The hot magmatic material ejected during an eruption can reach temperatures of 800 to 1000° Celsius. Increased thermal radiation can indicate the beginnings of an earth spasm. Earth-orbiting satellites make these monitoring tasks far easier than in previous years. Equipped with heat-sensory and magnetic field monitoring instruments, these satellites give man an upper hand in watching for potential natural catastrophes. Through observation, learning the component causal factors of various earth phenomena, and the constructive use of computer technology, man may learn to accurately forecast areas of trouble before trouble occurs.

One man who accurately predicted volcanic eruptions was Thomas A. Jaggar, former Director of the Hawaiian Volcano Observatory. A pioneer in the study of volcanology, Jaggar studied the Hawaiian volcanoes Mauna Loa and Kilauea. He believed that their activity was rhythmic, and occurred in certain divisible cycles. Using the period between 1790 and 1924 for his analyses, Jaggar broke down the volcanic activity of this 132-year supercycle into two 66-year cycles, four cycles of 33 years, and 12 cycles of approximately 11 years. Jaggar found this 11-year cycle to correspond with the 11 1-year cycle of sunspots—periods of volcanic activity seemed associated with maximal sunspot activity, while periods of quiescence related to miminal sunspot activity. The solar influences on volcanic activity and the possible effects of solar forces on other earth phenomena will be explored in the following chapters.

The possibility of cyclical phases in volcanic activity has led to the study of volcanoes and their individual cycles of volcanic activity as a way to signal warning of potential eruption. Studies of recorded volcanic activity between 1669 and 1928 at Mount· Etna, Europe's largest volcano, are said to show fluctuations in the cyclical phases: from 1775 to 1809, Etna erupted at nine-year intervals; 1809 to 1865, six-year intervals; 1865 to 1908, seven-year intervals. In studying Stromboli's activity, researchers have suggested a correlation between earth tides and volcanic eruptions. The theoretically ideal situation is the probability of eruption being greatest at the maximal amplitude of earth tidal action during its fortnightly cycle. Volcanoes, however, defy any attempt at generalization, and within the concept of an association between volcanic eruption and earth tides, a wide latitude is allowed for variables according to the geological features of the area and the composition of lava ejected from the volcano, whether basaltic or andesitic.

The history of Lassen Peak in northern California indicates a 65-year cycle of volcanic activity. Lassen Peak is at the southernmost end of the Cascade mountain range. Part of the Pacific Ring of Fire, the Cascades stretch about 435 kilometers in a north-south direction from Mount Garibaldi in British Columbia down to Lassen Peak. Fifteen major volcanoes are part of the Cascade Range. Until recently, the most active of them was Lassen Peak, which erupted in the winter of 1850-1851 and again in 1914-1915. If a 65-year cycle of volcanic activity at Lassen Park is indeed valid, then 1980 loomed as a potentially eruptive period in that area.

In March of 1980, one of the Cascade volcanic mountains came to life. Mt. St. Helens in Washington State had last erupted in 1856. It was showing signs of reawakening with an earthquake which reached a magnitude of 4. A week later, it announced itself with a minor eruption. For the next six weeks, Mt. St. Helens was the scene of moderate eruptions and moderate earthquake activity. Scientists noted a bulge on the north flank of the mountain. The bulge was continuously monitored, and showed a constant expansion of 1.5 meters per day. Earthquake activity was also monitored, and showed a decrease in the number of quakes, but an increase in their intensity. The scientists monitoring the emission of sulfur dioxide recorded about 30 tons per day, an insignificant amount compared to the 1000 tons per day emitted by active volcanoes. Scientists watched and waited for some forewarning of a major eruption. The warning never came. On May 18, 1980, the top of Mt. St. Helens blew off, changing the skyline of the Pacific coast and the landscape as far east as Montana. It was the week the sky fell.

Mt. St. Helens exploded with a force calculated at 10 megatons of TNT, or roughly 500 times as powerful as the atomic bomb dropped on Hiroshima during the Second World War. Triggered by an earthquake of a magnitude of about 5.0, the volcanic eruption of Mt. St. Helens blasted away 396 meters of the north slope of the 2950-meter volcanic mountain. Ash, rock, and gas were thrown from the volcano. Some of the debris penetrated the stratosphere, and reached heights of 19 kilometers. Scientists at the U.S. Geological Survey estimated 2.5 cubic kilometers of matter was ejected during the eruption. The toll of human casualties (30 dead, 34 missing) was relatively small for the size of the explosion. The May 18th eruption of Mt. St. Helens is calculated to have been the largest eruption in the area in the past 4000 years, and possibly the strongest volcanic eruption in the twentieth century.

Although Mt. St. Helens quieted down, the repose was short-lived. One week later, on May 25th, Mt. St. Helens experienced another major eruption. Other major eruptions of Mt. St. Helens occurred throughout 1980. Although the volcano has again quieted down, volcanic activity at Mt. St. Helens or at other volcanic mountains of the Cascade Range is not over. When Mt. St. Helens erupted in 1842, its volcanic activity continued sporadically for 15 years, ending in 1857. Mt. St. Helens may continue to erupt sporadically for many years.

An ominous sidelight to the volcanic activity in the Cascade range is the fact that when one volcano becomes active, other volcanoes in the Cascades become active. During the 15-year period between 1842 and 1857, when Mt. St. Helens was last active, eruptions occurred at Mt. Rainier, Lassen Peak, Mt. Baker, and Mt. Hood. In 1975, Mt. Baker gave signs of increasing activity. In July of 1980, Mt. Hood began to experience minor seismic activity. Observations of the Cascade range by scientists indicate the possibility of volcanic activity in the near future at four of the Cascade mountains: Mt. Hood, Mt. Baker, Mt. Rainier, and Lassen Peak.

The study of volcanoes is in its infancy, but once man begins to realize the impact of volcanic activity on his life conditions, he can use his technology to monitor important variables and be forewarned of incipient eruptions. Such knowledge can prevent the direct loss of lives. The indirect effects of eruptions may not be as easily remedied, although defusing or harnessing volcanic activity would go a long way in that direction.

The first known attempt to control lava flow from a volcanic eruption occurred during the 1669 eruption of Mount Etna. The flow from that eruption threatened the town of Catania some 10 miles (16 kilometers) away, a city destroyed by an earlier eruption in 122 A.D. As the lava stream approached the town, a man named Diego Pappalardo organized his fellow citizens to divert the flow from the town. Protected from the scorching heat by wet cowhides, Pappalardo, along with about fifty others, broke a channel through the lava flow to direct it away from Catania. Unfortunately, their efforts sent the lava stream in the direction of Paterno. When the townspeople of Paterno heard of this first attempt to divert lava flow, they put a stop to the experiment. Fearful of having their homes and fields destroyed, they sent out armed men to rout the Catanians. The lava flow resumed its original direction and destroyed much of Catania.

Later attempts to control lava flow have had mixed results. One success occurred with Kelut volcano on Java. The volcanoes of the East Indies are known for the devastating mud flows that accompany their eruptions. These mud flows are called *lahars* in Indonesian. Kelut volcano has a deep crater lake that erupts with flash floods of *lahars*. In 1919, an eruption sent *lahars* sweeping through 100 towns, killing 5000 people. To prevent such a catastrophe from occurring again, the Dutch colonial officials authorized the digging of a tunnel system to drain off the crater lake. The tunnels proved successful, for the lake stayed fairly well drained. In 1951, Kelut erupted again. While the eruption caused no *lahars*, it deepened the crater lake and the accompanying earthquakes wrecked the drainage system. There was a meager attempt to repair it, but the effort was futile. During Kelut's eruption in 1966, several hundred people were killed by the accompanying *lahars*.

Thomas Jaggar at the Hawaiian Volcano Observatory proposed another way of controlling lava flow. Jaggar's intimate knowledge of Kilauea and Mauna Loa allowed him to recognize the patterns of lava flow in their eruptions. From the records of 1868 to 1926, Jaggar saw a pattern of eruption moving up the mountain along a southwest rift. In March 1934, he predicted an eruption within two years along the northern flank of Mauna Loa, and he believed the direction of lava flow would endanger the city of Hilo, about 20 miles (32 kilometers) from the volcano. On November 21, 1935, Mauna Loa erupted, as foreseen, along the northeast rift. On December 22, the lava flow turned toward Hilo and, by the 26th, had advanced one-quarter the distance to the city. Back in 1931, Jaggar had proposed the idea of dropping bombs on lava flows to divert them. Jaggar believed that by creating new channels

100

for the lava flow, he could succeed at what the Catanians had first set out to do those many years earlier. On December 27, 1935, Jaggar saw his theory tested. From a height of 3500 feet (1067 meters) the U.S. Army Air Corps dropped twenty 600-pound bombs on two targets. The levees were breached, and the lava flowed in new directions. The city of Hilo was spared. Jaggar's proposal was a success. In 1942, the hypothesis was tested again at Mauna Loa, and again it proved successful. Other means of diverting lava flow are the building of dams or walls. This technique may be effective if the lava is basaltic, which is more liquid and tends to flow around obstacles. Viscous lava flow acts like a bulldozer, clearing everything from its path. An interesting by-product to the bombing tests in Hawaii is the knowledge that lava flows are far more sensitive to disturbances than had earlier been assumed. This information, and increased knowledge of volcanic activity, should lead us to consider, test, and effect other means of controlling the lava flow during a volcanic eruption.

In the meantime, we should prepare ourselves for increased volcanic activity. We have seen the worldwide consequences of volcanic eruptions. We have noted scientific predictions of increasing volcanic activity in the future. We are now more aware of the earth dynamics behind volcanicity. Our knowledge may not allow us to control earth spasms and render them harmless, but we can learn to live with their reality and mitigate their destructive potential. To counter the effects of volcanic activity, we should begin a program of food storage and seek channels for a more equitable distribution of food among the world's people. While severe calamities often appeal to man's compassion and generosity, the equitable distributions of food during tranquil conditions or in preparation for future need demands emotional maturity and a sense of justice. To date, attempts to develop such programs have often failed. However, there is a definite need for man to plan some program for food storage. The effects of Tambora, Krakatoa, Laki, Santorini, and countless other eruptions can happen today, and may occur in the near future.

VII

EARTH SPASMS

R: Did you hear? They had another earthquake in California last week. It was really traumatic.

L: I'm sure it was.

R: Two hundred and twenty-five people died.

L: And I imagine it must be pretty terrible for the thousands that survived.

R: I suppose you're right. They will suffer psychologically from the quake. But I hadn't thought of counting them, at first, as a way of measuring the extent of the disaster.

L: Most people don't. And yet you used the word "traumatic," which usually refers to a psychological effect.

R: You're right. Perhaps I was being careless. I was using it more to describe the earthquake itself than the effect of the quake on people.

L: But you were quite right. "Trauma" was first of all a surgical term. It's a wound which has been caused by a violent shock, a rupturing of the protective layer of skin. It also usually implied the repercussion, or reaction, on the part of the body that has been wounded.

R: Then it describes perfectly what an earthquake is—an internal wound within the earth which may eventually erupt with catastrophic effects. But how did this word, which describes something physical and concrete, also come to have a psychological sense?

L: Because the parallels are exact. A psychological trauma is also a physical or psychical shock which is felt to be life-endangering, and the wound which results is in the nervous system. Both kinds of trauma also involve a latency period which precedes the eruption of the symptoms.

R: A latency period?

L: Yes, a period of incubation during which the symptoms are invisible, because the wound is not felt.

R: Then an earthquake is both the cause and effect of a traumatic shock. What a confusion language creates by naming both traumas by the same word.

L: I don't think that it is language that creates that problem.

R: Then what does?

L: People. People who cause the wounding of the earth and then are shocked and confused by its traumatic reactions.

R: But what about the upheavals in the earth that occur independently of man's actions?

Time is the great oxidizer, creating major upheavals in the earth's crust. Each day our earth releases pent-up energies by means of spasms in the form of earthquakes. Although over a million earthquakes occur each year, with an average of two earthquakes each minute, an average of 19 such events a year can be classified as major quakes. In the past, major earthquakes have wreaked havoc and destruction rarely matched by any other form of earth spasm. Historical records provide evidence of two earthquakes, each of which claimed the lives of over three-quarters of a million people. Based on human casualties, the worst earthquake occurred in January 1556 in the Shensi province of northern China. A total of 830,000 people died. The second worst earthquake took place on July 28, 1976. This quake demolished the industrial city of Tangshan in China and claimed an estimated 750,000 lives. However, as Table 2 shows, China is not alone in suffering the severe effects of massive earthquakes. Earthquakes are a natural part of the past, present, and future; man must prepare himself for the potential devastation of future earthquakes.

In Chapter V we discussed at length the recent understanding of the mobility of earth dynamics. The scientific community no longer accepts the permanency of the continents and oceans. The theory of plate tectonics is now established fact. The earth's surface is a mosaic of lithospheric plates riding along the plastic flow of the earth's asthenosphere. The movements of these lithospheric plates create stress between two adjoining plates, with strain gradually accumulating along their contiguous borders. When the stress between the two plates reaches the breaking point, there is a sudden, abrupt release of the strain in the form of volcanic or earthquake activity. Like volcanoes, the majority of earthquakes occur along the borders of the lithospheric plates.

There are two major belts of earthquake activity in the world. One is the circum-Pacific belt, which extends around the Pacific Ocean from Chile along the west coasts of South and North America, the Aleutian islands, Japan, the Philippines, Indonesia, and New Zealand to New Guinea. The second major belt is the Alpide belt of Europe and Asia, which extends from the Azores through the Mediterranean, the Middle East, the Himalayas, and through Asia to Indonesia and New Guinea. About 80% of all earthquakes occur along the circum-Pacific belt, 15% along the Alpide belt, and the remaining five percent along the mid-oceanic ridges and elsewhere. While most of the world's earthquakes occur along the well-defined zones of lithospheric plate borders known to be seismically active, they have also occurred in unexpected regions. Past earthquakes have devastated areas located within the plates. In the United States, two examples of such intraplate quakes are the 1886 earthquake in Charleston, South Carolina, and the series of tremors in 1811-1812 centered around New Madrid, Missouri.

On the evening of August 31, 1886, an earthquake unexpectedly struck the city of Charleston, South Carolina, an area previously considered immune to earthquake risk. To the present day, this earthquake was the worst to hit the east coast of the United States. Much of Charleston was destroyed, with damage to an estimated 90% of the city's buildings, and at least 60 people dead. Aftershocks of decreasing intensity continued through September. The tremors from the Charleston earthquake were felt over a 5,180,000 square kilometer area, extending from Cuba to Boston and from Bermuda to Chicago.

The earthquakes around New Madrid, Missouri, during the winter of 1811-1812 are believed to have been the most powerful earthquakes ever recorded for the North American continent. This series lasted over a year, although the three major shocks took place within a two-month period. The first principal shock came at 2 A.M. on December 16, 1811, the second on January 23, 1812, and the third, and largest, on February 7, 1812. It's estimated that the magnitudes of these three major quakes reached 7.5, 7.3, and 7.8 on the Richter scale. Because of the low population density of the region, the death rate was low. However, this series of quakes caused land deformation that altered about 78,000 to 130,000 square kilometers of the topography of the Mississippi Valley region. The surrounding areas of southeastern Missouri, southern Illinois, southwestern Kentucky, northwestern Tennessee, and northeastern Arkansas were devastated. Visible waves rolled across the earth surface. Areas of land were uplifted 4½ to 6 meters, while other areas dropped 1½ to 3 meters. Swamps became dry land. Dry land became swamps or lakes. Two lakes formed by the New Madrid earthquakes are Lake St. Francis in eastern Arkansas (60 to 64 kilometers long, ¾ kilometer wide) and Reelfoot Lake in northwestern Tennessee (13 to 16 kilometers long, three to five kilometers wide). The course of the Mississippi River was altered. The major shocks at New Madrid were felt over a 5,180,000 square kilometer area, as far away as the Atlantic coast, Canada, the Rocky Mountains, and the Gulf of Mexico.

Although the cause of such intraplate earthquakes still defies complete understanding, they are associated with faults—areas of rock fracture where the two sides have been displaced in relation to each other and in a direction parallel to the fracture. Movement along the faults can be vertical (a dip-slip fault), horizontal (a strike-slip fault), or a combination of vertical and horizontal movement. While slow-slip or creeping movement along a fault may go undetected, pressure eventually builds up until there is a sudden release of the accumulated strain and a consequent earthquake. From his study of the displacement along the San Andreas fault after the 1906 San Francisco earthquake, Harry Fielding Reid of Johns Hopkins University proposed, in 1910, the elastic rebound theory of earthquake generation. This theory holds that rocks are elastic, and, therefore, store up frictional energy from the stress of movement along the fault until they reach their breaking point. Like an overwound spring, they break at their weakest point and snap back to an unstrained state. The elastic rebound is the snapping-back process when the elastic energy is suddenly released, generating heat and elastic waves. These elastic waves are the seismic waves of earthquakes.

In the mid-1930's, Beno Gutenberg and Charles Richter, both of the California Institute of Technology, divided earthquakes into three major categories, based upon their depth of focus: (1) normal or shallow-focus earthquakes, where the focus lies between 0 and 70 kilometers; (2) intermediate-focus earthquakes, where the focus lies between 70 and 300 kilometers; (3) deep-focus earthquakes, where the focus lies between 300 and 700 kilometers. Reid's elastic rebound theory is accepted as valid for shallow-focus and intermediate-focus quakes. Deep-focus earthquakes are associated with ocean trenches, where a lithospheric plate descends into the asthenosphere. As a result of this descent, the epicenters of deep-focus earthquakes are generally found to lie along a plane with an angle to the horizontal of an average 45°. Known as Benioff zones, after Hugo Benioff of the California Institute of Technology, who identified this feature of ocean trenches and their seismic activity, these sloping seismic zones tend to dip down toward the continental edge of the lithospheric plate which is being underridden. Roughly 90% of deep-focus earthquakes occur along the circum-Pacific belt.

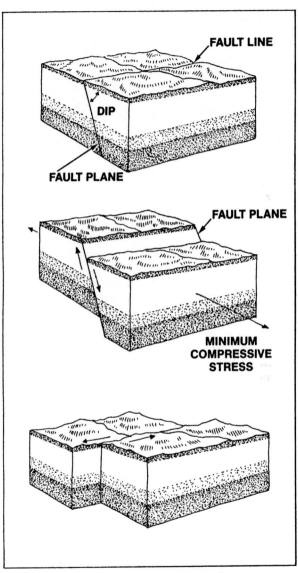

How land masses interact during an earthquake.

107

At the other extreme of focus depth, shallow-focus earth-quakes account for over 75% of earthquake energy released, and tend to be the most destructive of all tremors. These earth-quakes are associated with the upwelling along the ocean ridge systems and with movements along the faults. All the earth-quakes known to occur along the San Andreas fault in Califor-nia have been shallow-focus. These continually confirm the theory of elastic rebound as the mechanism for shallow-focus earthquakes.

While we recognize earthquake activity to be a natural consequence of the creation or destruction of lithospheric plates, and the release of built-up energy or stress, some scientists propose other causes of earthquake activity. One such proposal asserts a correlation between the Chandler wobble and earth-quakes.

As we noted in our discussion of continental drift in Chap-ter V, the earth acts like a bar magnet. This imaginary magnet is located at the earth's center, and lines up in the same general direction as the earth's axis of rotation, but with a slight vari-ance. Both surrounding and permeating the earth, the earth's magnetic field is in a continual state of flux. Its strength and positions are constantly changing. Because the earth's magnet-ic axis and its geographical axis do not coincide, compass readings have to be corrected for the deviation between the two. In 1891, Seth Carlo Chandler investigated the recorded variations in latitude. He believed the variations to result from the earth's wobbling motion, which he attributed to two cycles—an annual cycle and a 428-day, or 14-month cycle. Chandler believed that the annual cycle was the consequence of seasonal fluctuations in the oceans and the atmosphere. The second cycle, now called the Chandler wobble, still eludes complete comprehension. One explanation offered for it is the elasticity of the earth's interior. However, the elasticity of the earth and other rotational forces would be expected to gradually modify the wobble, and the Chandler motion would decrease to half its amplitude in 14 years. As this does not occur, there must be influences that regenerate the wobble.

Cosmic factors account in part for this wobble. In the ionosphere, the earth's upper atmosphere, the radiation of the sun strips electrons from the oxygen and nitrogen atoms. The position and negative charge particles of the ionosphere (the ions and electrons) make the air an electrical conductor, and the electric currents generated in the atmosphere create magnetic fields which contribute to variations in the earth's magnetic field. In the recent work by Walter Munk and E.S.M. Hassan, both of the University of California at San Diego, the annual variations in the magnetic field were largely attributed to seasonal changes in the distribution of the atmosphere. Confirming the 1901 suggestion of Rudolf Spitaler of the University of Prague, Munk and Hassan found that changing winds and atmospheric conditions are the primary factors in the annual variations. However, Munk and Hassan also calculated that atmospheric influences on the Chandler wobble are $\frac{1}{10}$th to $\frac{1}{100}$th times too small to account for the observed motion. Some factor other than atmospheric conditions had to contribute to the excitation of the Chandler wobble.

In 1906, John Milne proposed a connection between earthquakes and the Chandler wobble. Evidence suggests a relationship between the two. Frank Press, Presidential Science Advisor and Massachusetts Institute of Technology professor, argued in the mid-1960's that displacement during a large earthquake is much greater than originally thought, sometimes extending to several thousand kilometers from the quake's epicenter. Press based his argument on distant strain measurements and the theoretical predictions of elasticity theory. Lalantendu Mansinha of the University of Western Ontario, Canada, and Douglas Smylie of York University, Canada, applied the elasticity theory and Press's notion of the more extensive displacement from earthquakes to calculate the excitation to the Chandler wobble by earthquakes. Mansinha and Smylie published their discovery in *Science* in 1968: large earthquakes could indeed cause excitation to the Chandler wobble and could affect movement. Comparing the dates of 22 major earthquakes with the dates of unusual changes in the magnetic pole's path, Mansinha and Smylie found a correlation for 15 out of the 22 quakes. One interesting and unexpected result of their study was the discovery that changes in the magnetic pole's path often preceded the correlated earthquake.

The two cycles that influence the earth's wobble sometimes work against each other, and polar motion is minimal. Every seven years the two cycles coincide and work together. At such times, the polar drift increases its rate of motion, and there appears to be a high incidence of large magnitude earthquakes. Surveying the recorded earthquakes during the twentieth century up to 1970, Charles A. Whitten of the University of California at Los Angeles verified this correlation between pronounced polar motion and energy release by earthquakes. If Whitten's observations of the peaks in the seven-year cycle of polar motion holds true for the future, then 1985, 1992, and 1999 should be years of intense earthquake activity.

Although the correlation between increased polar motion and earthquake activity is readily apparent, scientists are perplexed by the old "chicken and the egg" problem. Which of these two phenomena causes the other? Do earthquakes aggravate the wobble? Or do polar variations trigger earthquakes? While the problem is not yet solved, it seems that cause and effect operate both ways.

Don L. Anderson, director of the Seismological Laboratory at the California Institute of Technology, has proposed a connection between explosive volcanism, atmospheric conditions, and the earth's wobble that triggers earthquake activity. Increased volcanic activity projects more particulate matter into the atmosphere, which reduces the solar radiation reaching the earth's surface. Changes in the incoming solar radiation alter global wind patterns, and thereby affect the length of the day. From his observations, Anderson found an increase in seismic activity during the upswings in the length of the day at the turn of the twentieth century. As noted in the work of Munk and Hassan, changing wind patterns and atmospheric conditions are seen to excite the earth's wobble. Because of upswings in the length of the day, which started around 1960, it is likely that earthquake activity will be on the rise in the future. This increase may well be aggravated by man's actions of air pollution and weather modification, subjects we will consider at length in following chapters.

Man's awareness of the effects of solar radiation on atmospheric conditions, the earth's wobble, and seismic activity, calls for his reevaluation of cosmic factors and how they influence conditions like earthquake generation. A likely catalyst of earthquake activity is the gravitational forces or tidal actions. Based on records dating from 231 B.C. for the Chinese provinces of Hopei and Shansi, a recent Chinese study found a correlation between earthquake activity and tidal forces. In

Hopei province, statistics indicate that 75% of large magnitude earthquakes occur within two days of the sun's and moon's aligning with the earth, or their being at a 90° angle to each other. At such times, the forces of the sun and the moon acting together tend to pull the land area of Hopei outwards. In Shansi province, however, most earthquakes occur when the sun and moon are not aligned. The Chinese study attributes dissimilarity in earthquake generation in these two provinces to geological differences. The distinctive formations of each area will determine whether earthquakes are triggered by the outward pull or by the downward pressure of tidal forces.

A more controversial theory considers all the planets of our solar system as playing a significant role in earthquake generation. Babylonian tradition considered Saturn, Jupiter, and Mars as responsible for earthquakes. Although modern skeptics deride the belief of planetary influences on earthly conditions as remnants of ancient superstitions, man's increasing knowledge of his universal environment is reawakening him to the reality of a gestalt of our solar system. We have already noted the findings of John H. Nelson that two or more planets forming a 0°, 90°, or 180° alignment with the sun seem to intensify magnetic storms on the sun, thereby interfering with radio transmissions on earth. We also mentioned the correlation found by Rudolph Tomaschek between planetary configurations and major earthquakes in his analysis of 134 earthquakes with magnitudes above 7.7 on the Richter scale. In their popular book, *The Jupiter Effect,* John Gribbin and Stephen Plagemann also consider the possible correlation between planetary alignment and earthquake generation.

In Chapter II, we noted that the gravitational forces of the planets in our solar system exert tidal actions on the sun. Gribbin and Plagemann believe these tidal forces reach their maximum every 179 years, when all nine planets of the solar system line up on the same side of the sun. The next time this planetary alignment occurs will be during 1982 to 1984, at which time the 11-year sunspot cycle will also reach its peak. According to Gribbin and Plagemann's scenario, gigantic magnetic storms on the sun are likely to result. Bursts of solar flares will affect the earth's atmosphere and disrupt weather patterns. This change will intensify the earth's wobbling motion, which in turn will trigger off earthquakes throughout the world. One area Gribbin and Plagemann believe to be particularly prone to a major earthquake at that time is the San Andreas fault in California. Due to the absence of increased seismic activity recorded during the years of 1445, 1624, and 1803, when similar planetary alignments occurred, some scientists reject this theory of planetary alignment and earthquake generation. However, before we dismiss such a theory, we should recognize that any phase of cyclical activity never contains the identical coordinates of the same phase in an earlier or later cycle. Nature is a process of alternating, rhythmic behavior—of cycles within cycles.

Although man at present does not totally understand the causes of earthquake generation, he is well aware of their effects. In order to measure the effects of earthquakes, two types of scales are used to describe an earthquake. One measures the intensity of the quake, the *earthquake's effects* on the human conditions and the geology of a given area. The other scale measures the magnitude of the quake, the amount of *energy released* by the earthquake.

Measuring the effects in a given area, intensity scales are dependent upon such variables as: (1) local geology, (2) distance of the area from the epicenter of the quake, (3) population density, (4) concentration and type of building structures, and (5) time of day of occurrence. In 1883, Michele De Rossi and Alphonse Forel devised a scale with a range of 1 to 10 to describe the intensity of an earthquake. The De Rossi-Forel scale was replaced in 1902 by one devised by Giuseppi Mercalli with a range of I to XII. In 1923, A. Sieberg refined the Mercalli scale, and in 1931 H.O. Wood and Frank Neumann adapted it to North American conditions. Known as the Modified Mercalli Scale or the Wood-Neumann Scale, it was reworked by Charles F. Richter in 1956. Richter's 1956 version of the Modified Mercalli Scale (given in Table 1) is now generally accepted in measuring the effects of an earthquake in a given area.

TABLE 1
MODIFIED MERCALLI INTENSITY SCALE
(1956 version*)

Masonry A, B, C, D. To avoid ambiguity of language, the quality of masonry, brick or otherwise, is specified by the following lettering.

Masonry A. Good workmanship, mortar and design; reinforced, especially laterally, and bound together by using steel, concrete, etc.; designed to resist lateral forces.

Masonry B. Good workmanship and mortar; reinforced, but not designed in detail to resist lateral forces.

Masonry C. Ordinary workmanship and mortar; no extreme weaknesses like failing to tie in at corners, but neither reinforced nor designed against horizontal forces.

Masonry D. Weak materials, such as adobe or poor mortar; low standards of workmanship; weak horizontally.

Intensity Value	Description
I.	Not felt. Marginal and long-period effects of large earthquakes.
II.	Felt by persons at rest, on upper floors, or favorably placed.
III.	Felt indoors. Hanging objects swing. Vibration like passing of light trucks. Duration estimated. May not be recognized as an earthquake.
IV.	Hanging objects swing. Vibration like passing of heavy trucks, or sensation of a jolt like a heavy ball striking the walls. Standing motor cars rock. Windows, dishes, doors rattle. Glasses clink. Crockery clashes. In the upper range of IV wooden walls and frame creak.
V.	Felt outdoors; direction estimated. Sleepers wakened. Liquids disturbed, some spilled.

(*Original 1931 version by H.O. Wood and Frank Neumann abridged and rewritten in 1956 by Charles F. Richter, as given in Richter's *Elementary Seismology*, pp. 136-138, W.H. Freeman & Co., 1958.)

113

Small unstable objects displaced or upset. Doors swing, close, open. Shutters, pictures move. Pendulum clocks stop, start, change rate.

VI. Felt by all. Many frightened and run outdoors. Persons walk unsteadily. Windows, dishes, glassware broken. Knickknacks, books, etc., off shelves. Pictures off walls. Furniture moved or overturned. Weak plaster and masonry D cracked. Small bells ring (church, school). Trees, bushes shaken visibly, or heard to rustle.

VII. Difficult to stand. Noticed by drivers of motor cars. Hanging objects quiver. Furniture broken. Damage to masonry D, including cracks. Weak chimneys broken at roof line. Fall of plaster, loose bricks, stones, tiles, cornices, also unbraced parapets and architectural ornaments. Some cracks in masonry C. Waves on ponds; water turbid with mud. Small slides and caving in along sand or gravel banks. Large bells ring. Concrete irrigation ditches damaged.

VIII. Steering of motor cars affected. Damage to masonry C; partial collapse. Some damage to masonry B; none to masonry A. Fall of stucco and some masonry walls. Twisting, fall of chimneys, factory stacks, monuments, towers, elevated tanks. Frame houses moved on foundations if not bolted down; loose panel walls thrown out. Decayed piling broken off. Branches broken from trees. Changes in flow or temperature of springs and wells. Cracks in wet ground and on steep slopes.

IX. General panic. Masonry D destroyed; masonry C heavily damaged, sometimes with complete collapse; masonry B seriously damaged. General damage to foundations. Frame structures, if not bolted, shifted off foundations.

Intensity Value	Description

	Frames cracked. Serious damage to reservoirs. Underground pipes broken. Conspicuous cracks in ground. In alluviated areas sand and mud ejected, earthquake fountains, sand craters.
X.	Most masonry and frame structures destroyed with their foundations. Some well-built wooden structures and bridges destroyed. Serious damage to dams, dikes, embankments. Large landslides. Water thrown on banks of canals, rivers, lakes, etc. Sand and mud shifted horizontally on beaches and flat land. Rails bent slightly.
XI.	Rails bent greatly. Underground pipelines completely out of service.
XII.	Damage nearly total. Large rock masses displaced. Lines of sight and level distorted. Objects thrown into the air.

The intensity scale measures the effects of an earthquake on a local area. The magnitude scale measures the amount of energy released. Commonly known as the Richter scale, the magnitude scale was developed by Richter and Beno Gutenberg in the mid-1930's and measures the amplitude of surface waves produced by the earthquake. Ranging from 0 to 8.6 (although variations of the range exist), this scale is logarithmic. A whole number on this scale is a jump of 10 times in the amplitude of surface waves or ground motion. For example, a magnitude 7.0 earthquake is 10 times stronger than an earthquake of magnitude 6.0. An earthquake of magnitude 7.0 is 10 x 10 x 10 x 10 (or 10,000 times) stronger than a magnitude 3.0 earthquake. The ground motion is associated with the energy released during an earthquake, but each increase of a whole number in magnitude relates to a jump of about 30 times the energy released. In our example of the difference between a magnitude 7.0 earthquake and an earthquake of magnitude 3.0, the larger quake will release approximately 810,000 times (30 x 30 x 30 x 30) the energy released in the smaller quake.

Just as the intensity scales were revised there's been a recent suggestion to revise the magnitude scale. Hiroo Kanamori of the California Institute of Technology has proposed and developed a magnitude scale based upon seismic moment. Instead of measuring the seismic energy released at the hypocenter or source of the earthquake, the seismic moment measures the energy released along the entire fault. While many scientists believe that this measurement reflects the true magnitude of an earthquake, the Richter scale remains the more widely used scale of magnitude at present.

While intensity scales are more qualitative or subjective in their measurement, magnitude scales are more quantitative or objective. For the magnitude scales, earthquake waves are recorded by a seismograph. The first instrument capable of detecting an earthquake at a distance, was invented by the Chinese scholar Chang Heng in 132 A.D. The seismograph was invented by John Milne in the late nineteenth century. The instrument has a pendulum or spring-suspended weight placed in a rigid frame which swings when an earthquake wave passes through the earth underneath it. As accurate timetables for the travel times of P and S waves (see Chapter IV) for distances up to 11,265 kilometers have been calculated from repeated measurements, the distance of the earthquake epicenter from the seismograph station can be found from the differences in travel times for the P and S waves. Although the distance between the seismograph station and the earthquake location can be calculated by using one station, the exact location of the epicenter is found by a process of triangulation, which employs at least three seismograph stations. There are over 1000 seismograph stations in the world. One of the most impressive centers for the study of seismic waves is the Large Aperture Seismic Array (LASA). Built in 1965 near Billings, Montana, LASA has more than 500 seismometers in 21 clusters spread over an area of 200 kilometers.

In order to familiarize ourselves with the process of earthquakes, we have considered the abstract concepts of earthquake generation. However, for the layman, interest in earthquakes primarily centers on the effects they have upon us and our life conditions. As Table 2 indicates, earthquakes have taken heavy tolls of human life in the past. But Table 2 merely provides statistics. To flesh these out, we turn our attention to various examples of severe earthquake activity and the effects they have wrought upon human conditions.

TABLE 2
SEVERE EARTHQUAKES OF THE PAST
(Based on human casualty estimates
of 5000 deaths or more)

YEAR	LOCATION	ESTIMATED DEATHS
342	Turkey	40,000
565	Turkey	30,000
856	Greece	45,000
1038	China	23,000
1057	China	25,000
1268	Asia Minor	60,000
1290	China	100,000
1293	Japan	30,000
1456	Naples	30,000
1531	Portugal	30,000
1556	China	830,000
1641	Persia	30,000
1667	Caucasia	80,000
1693	Italy	60,000
1716	Algeria	20,000
1737	India	300,000
1755	Persia	40,000
1755	Portugal	70,000
1759	Lebanon	20,000
1783	Italy	50,000
1797	Ecuador	40,000
1822	Asia Minor	22,000
1828	Japan	30,000
1868	Peru & Bolivia	25,000
1868	Ecuador & Colombia	70,000
1891	Japan	7,000
1896	Japan	22,000

YEAR	LOCATION	ESTIMATED DEATHS
1905	India	20,000
1908	Italy	100,000
1915	Italy	30,000
1920	China	180,000
1923	Japan	140,000
1927	China	200,000
1932	China	70,000
1935	India	60,000
1939	Chile	30,000
1939	Turkey	23,000
1944	Argentina	5,000
1948	Japan	5,200
1949	Ecuador	6,000
1960	Morocco	12,000
1960	Chile	5,700
1962	Iran	12,000
1968	Iran	12,000
1970	Peru	70,000
1972	Iran	5,400
1972	Nicaragua	5,000
1976	Guatemala	23,000
1976	China	750,000
1978	Iran	25,000
1980	Algeria	10,000

As we noted at the beginning of this chapter, China has suffered the two worst earthquakes in recorded history. In January 1556, the province of Shensi was devastated by a quake with an estimated magnitude of between 8.0 and 8.3 on the Richter scale. It destroyed an area extending 870 kilometers and caused the deaths of at least 830,000 people. The second worst earthquake in history occurred on July 28, 1976, in Tangshan, 160 kilometers southeast of Peking. The quake's initial shock registered 8.2 on the Richter scale, and the second tremor, 7.9 magnitude. Before the quake, Tangshan had a population of one million, and its extensive coal-mining operations made it the third most important industrial area in China. Although information on this quake is limited, it is now estimated that some 750,000 people lost their lives in it. Accounts tell that citizens of Tangshan were awakened before the quake by an incandescent glow that lit up the predawn sky for hundreds of kilometers. When the quake hit, some people were hurled two meters into the air. Trees and vegetation were flattened. In some areas, the earth was ripped open. In other areas, craters pocked the land. During the following months, over 300 aftershocks kept the people in a perpetual state of alarm. Although China has been at the forefront of earthquake prediction, the Tangshan earthquake occurred without a specified warning. This happened despite Chinese scientists' being aware of preliminary omens of earthquake activity; they had, in fact, announced in early 1976 the probability of an impending quake in the Tangshan or Peking areas. Due to conflicting information, however, no specific prediction was made. Having catalogued their earthquakes for over 3000 years, the Chinese recorded close to 1000 destructive earthquakes, approximately 18 of them considered to have had magnitudes of 8.0 or greater. Chinese earthquakes are generally explained by the concept of plate tectonics. The Indian subcontinent crashes into Eurasia, and triggers the earthquakes in China.

With a magnitude estimated at between 8.75 and 9.0 on the Richter scale, one of the strongest earthquakes ever to occur struck Lisbon, Portugal, on November 1, 1755. This quake was felt over an area of about 3,240,000 square kilometers throughout Europe and in much of North Africa. It set lakes into oscillation as far away as Scotland and Scandinavia. The first shock struck between 9:30 and 10:00 A.M. This initial tremor was followed by two other violent shocks. As it was All Saints' Day, much of Lisbon's population of 275,000 was at worship in various churches throughout the city. Many were crushed by the collapsing roofs and walls. Some were consumed by the fires that broke out in the city. Still others perished from the *tsunamis* with heights of 15 meters that roared down on the city. Not only did the *tsunamis* lay waste to Lisbon, but these seismic sea waves generated by the quake also raced across the Atlantic at speeds of close to 875 kilometers an hour, and ravaged parts of the West Indies. In the wake

119

of this catastrophic earthquake, Lisbon was reduced to chaos. Half the city was destroyed, and at least 70,000 people were killed.

One earthquake that resulted in extensive damage to two major cities occurred in Sagami Bay off the coast of the main Japanese island of Honshu on September 1, 1923. Referred to as the Kwanto earthquake, this event was generated by a submarine landslide about 75 kilometers from the port city of Yokohama, and 100 kilometers from Tokyo. Although a *tsunami* pounded the Honshu coast, the worst damage from the quake was done by fire. The earthquake hit shortly before noon, while the people were readying their charcoal stoves for the midday meal. Knocked over by the shocks, the stoves set off fires throughout the Yokohama-Tokyo area. In both cities, firefighting equipment was destroyed in the quake, and the water mains were broken. Fanned by high winds, the fires leaped from one point to another. Survivors fled to open areas to avoid the flames. In one particularly grisly episode, close to 40,000 people congregated in an open area in Tokyo. As the fires consumed the surrounding buildings, it sucked the oxygen from the air, and all 40,000 suffocated. For two days the fires raged uncontrolled, until they burned themselves out. The earthquake and its effects destroyed about 75% of Yokohama and 60% of Tokyo. Half a million homes were lost. Although the official death toll from the Kwanto earthquake was 99,331 dead out of a population of two million, estimates run as high as 140,000 dead. During all of September, over 1,250 aftershocks were recorded on the Tokyo seismograph. In Sagami Bay, the ocean floor is believed to have been elevated in some areas by as much as 225 meters, while in other areas subsidence ranged to 180 meters.

Japan is particularly prone to earthquake activity, for it is located near the boundaries of the Pacific, Philippine, and Eurasian lithospheric plates. The Pacific and Philippine plates push against each other as they slide beneath the Eurasian plate. Consequently, the east coast of the island of Honshu is being pulled downward. This action is creating anxiety among many Japanese scientists, including Masakazu Ohtake of the Japanese National Research Center for Disaster Prevention, for there is a growing expectation of a major earthquake within the next 10 years in the Tokai region of central Japan, about 160 kilometers southwest of Tokyo. During the last 85 years, parts of the Tokai region have subsided 30 centimeters, and considerable stress in the earth's crust is evident. Other scientists fear a major earthquake will hit Tokyo in the near future. According to records of the last thousand years, Tokyo has been struck by a major earthquake on the average of once every 69 years. As two-thirds of these earthquakes occurred during the last 13 years of each cycle, we might expect a major earthquake for Tokyo between 1979 and 1992. Should such a major earthquake hit Tokyo in the future, the results could be devastating. Estimates claim a possible 560,000 deaths, and some range as high as three million deaths. Lest the reader feel that Tokyo should be classed as a doomed area, we should recognize that the prediction of an impending earthquake in Tokyo is based upon historical records. But unlike the Tokai region, geological evidence indicates that the growing tension in the Tokyo region has not yet reached a breaking point.

Avalanches caused by earthquake activity have also yielded grim results in the past. One example is the earthquake that struck Peru on May 31, 1970, with a magnitude of 7.7. Its center was in the Pacific, about 25 kilometers west of the port city of Chimbote and at a focal depth of about 43 kilometers beneath the ocean floor. An area of 65,000 square kilometers was devastated, and as many as 70,000 lives were lost. Although many died in the collapse of their poorly constructed dwellings, over 20,000 were killed in the avalanche of rock and ice dislodged from the northern peak of Huascarán, Peru's highest mountain. Estimates put the size of the avalanche at 915 meters wide, 1½ kilometers long, and containing approximately 2,265,000 kiloliters of material. Fluidized by the air trapped between the glacial surface and the moving mass, this avalanche is believed to have hurtled into the valley below at speeds of close to 450 kilometers an hour. The towns of Yungay and Ranrahirca lay buried in its path.

Ten years before the Peruvian quake of 1970, one of the strongest earthquakes in the twentieth century occurred along the same ocean trench but further south. Registering 8.5 on the Richter scale, the May 21, 1960, earthquake had a focal depth of about 65 kilometers and shook a large area of south central Chile. Its tremors set the world vibrating like the ringing of a bell for nearly two weeks. From the study of this earthquake, scientists were able to verify our present concept of the earth's structure. Two areas in Chile suffered the maximal intensity of the quake—the coastal region from Puerto Saavedra to Isla de Chiloé and inland in the Chilean Lakes region along the Reloncavi fault. The quake is also thought to have caused the eruption of Puyehue, a volcano dormant since 1905. Because of this quake, land areas in Chile were displaced by as much as 20 meters. Even in Los Angeles, California, there was ground displacement of a maximum two millimeters. More than 5700 people lost their lives, many of them from the *tsunamis* generated by the quake. Shortly after the main shock, the seas began to recede from the coast. About 15 minutes later, waves six meters in height crashed onto the Chilean shoreline and raced as far as three kilometers inland. When the news of the *tsunamis'* effects on Chile reached Hawaii, warnings were issued for the people living along the coast of the island to evacuate. The time of arrival for the *tsunamis* was calculated at close to 15 hours, an estimate accurate to within one minute of the actual time. Having travelled across the Pacific at speeds of up to 700 kilometers an hour, these *tsunamis* crashed into Hilo at heights of 10 meters. Some people refused to heed the warnings, which resulted in 62 deaths in Hawaii. From Hawaii, the *tsunamis* moved on to Japan, where their arrival time was also accurately predicted. Since Japan is 17,060 kilometers away from Chile, the potential for damage from these *tsunamis* was underestimated. In Japan, 180 people lost their lives.

Another massive earthquake, registering about 8.5 on the Richter scale, occurred along the circum-Pacific belt four years later. On the evening of Good Friday, March 27, 1964, northern Prince William Sound off Alaska was hit by a quake believed to have been 2000 times more powerful than any nuclear bomb ever exploded. During the next month, 19 aftershocks registered above 6.0 magnitude, and for the next two months the area was shaken by approximately 12,000 aftershocks. Although the death toll of 131 was minor for such a large quake, the geological changes were enormous. Displacement of the earth's crust occurred as far west as Hawaii. The quake transformed about 200,000 square kilometers of land, and the shock waves were felt throughout the world. The U.S. Capitol building in Washington, D.C., is said to have moved five centimeters. The earth moved up and down about 15 centimeters in Houston, Texas, and about eight millimeters in Iran. But Alaska suffered the brunt of vertical displacement. While the average displacement ranged between 1 and 2½ meters, the northern section of Montague Island rose between three to six meters, and the southern section rose 10 meters or more. In Anchorage, some downtown buildings sank six meters below street level. Landslides devastated parts of the city. On Kenai Peninsula, entire mountains moved laterally 1½ meters and sank by two meters. Atmospheric waves generated by the earth's movement caused water levels to fluctuate three to six meters in wells in Georgia, U.S.A., and fluctuations of water levels were also reported in wells in Europe, Asia, Africa, and Australia. Almost all of the deaths from this quake are attributed to the consequent *tsunamis*. With heights of six to 30 meters, they ravaged the Alaskan coastline. Kodiak, Seward, and Valdez suffered severe damage. Ruptured oil tanks ignited fuel fires on the water. Sixteen hundred kilometers to the south, the *tsunamis* were responsible for the deaths of 10 people in Crescent City, California. In San Francisco Bay, the water bobbed up and down one meter. Water surges caused boats tied up in Houston, Texas, and New Orleans, Louisiana, to break their moorings. Moving at estimated speeds of 800 kilometers an hour, the *tsunamis* led to evacuations in the coastal regions of Hawaii, Japan, and the Philippines.

Alaska is not the only prime target for earthquake activity in the United States. In the contiguous United States, California experiences the vast majority of all earthquakes, and is commonly referred to as "earthquake country." Indeed, California's high-density population makes it more vulnerable than Alaska to the catastrophic consequences of earthquakes upon human life conditions. California is a complex labyrinth of faults, the most notorious of which is the San Andreas fault. Between 1934 and 1977, more than 8,800 earthquakes of magnitude 4.0 or greater were recorded along the San Andreas fault. Extending about 1000 kilometers in a southeasterly-northwesterly direction, the San Andreas stretches from the Salton Sea through San Bernardino and 59 kilometers east of Los Angeles, up through the San Francisco area, and out to sea around Cape Mendocino in the northern part of the state. The San Andreas is a strike-slip fault, the type where the two opposing sides slide along each other. In the case of the San Andreas fault, the

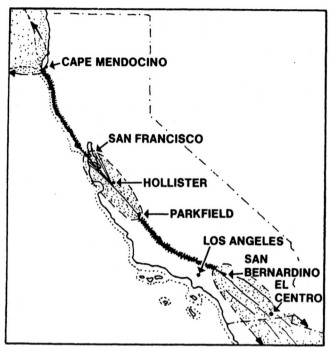

The San Andreas fault of California.

Pacific plate with the western edge of California, which includes Los Angeles, is sliding at an approximate rate of five to six centimeters a year in a northwesterly direction against the edge of the North American plate, on which San Francisco stands. In this case, the ground displacement tends to be horizontal; thus, the *tsunamis,* associated with vertical displacement, are unlikely to occur with California quakes.

One of the most interesting and perhaps paradoxical aspects of the San Andreas fault is the focal depth at which earthquakes occur. Although the San Andreas lies along the borders of the Pacific and North American lithospheric plates, earthquakes along this fault are of a shallow focus. Most California quakes occur at a focal depth of 10 to 20 kilometers below the earth's surface. An answer to this seeming paradox was suggested in the mid-1960's by the laboratory experiments on rocks subjected to high pressure by William Brace and his colleagues at the Massachusetts Institute of Technology. Brace proposes that earthquake-free motion or aseismic creep might occur along the fault at lower depths, due to the mineralogic changes of the rocks, changes in the pore pressure conditions of the rocks, and temperature increases at these lower depths which would then allow a self-lubricating process and greater plasticity of flow. We will see how these and other experiments have provided valuable information in considering the feasibility of modifying the potential violence of earthquake activity. However, the reality of the conditions shows California to be highly susceptible to major earthquake activity.

The two most powerful earthquakes to strike California occurred in 1857 at Fort Tejon and in 1906 in San Francisco. In terms of human casualties, however, the San Francisco earthquake was the most severe. It struck at 5:12 A.M. on April 18, 1906, along the San Andreas fault at a focal depth of about 15 kilometers and with an estimated magnitude of 8.2. The Pacific plate on the west side of the fault had moved with a lurch in a northwest direction to the North American plate on the east side of the fault. The offset along the fault was primarily horizontal, and the rupture extended 430 kilometers from San Juan Bautista in San Benito County to near Cape Mendocino in Humboldt County. While vertical displacement at its maximum was less than one meter, horizontal displacement along the fault averaged three meters, with a maximal displacement of seven meters north of San Francisco. The quake was felt over an area of 970,000 square kilometers. The ground was rent apart, landslides occurred, buildings collapsed, and fires raged out of control. San Francisco burned for three days, 500 city blocks were consumed, and 700 people lost their lives in the aftermath of this quake.

As Table 3 indicates, California has repeatedly experienced severe earthquake activity. In 1933, an earthquake caused extensive damage to the Los Angeles area cities of Long Beach, Compton, Torrance, and Garden Grove, claiming the lives of 120 people. At a little after 6:00 A.M. on February 9, 1971, an earthquake of 6.5 magnitude struck the San Fernando Valley, an unexpected location. This area had been seismically inactive for thousands of years. The death toll reached close to 65 people, but it could have been much worse. Since the quake hit in the early morning, most people were still at home. Had it happened during the rush hour, the death count would have been considerably higher. Although most of the deaths resulted from the collapses at the Veterans Hospital, the Van Norman Dam came close to giving way. Had it collapsed, residential areas would have been deluged, and the 1971 San Fernando earthquake might have been catastrophic.

TABLE 3
MAJOR CALIFORNIA EARTHQUAKES OF THE PAST

YEAR	LOCATION	ESTIMATED DEATHS
1812	San Juan Capistrano	40
1836	San Francisco	-
1838	San Francisco	-
1857	Fort Tejon	1
1865	San Francisco	-
1868	Hayward	30
1872	Owens Valley	27
1899	San Jacinto	6
1906	San Francisco	700
1915	Imperial Valley	6
1925	Santa Barbara	13
1933	Long Beach	120
1940	Imperial Valley	8
1941	Santa Barbara	-
1941	Torrance, Gardena	-
1952	Kern County	14
1957	San Francisco	-
1969	Santa Rosa	-
1971	San Fernando	65

While California has been spared the massive human destruction from earthquakes suffered in other countries, it may not be so lucky in the future. Recent observations show some sections of the San Andreas fault with an accelerated rate of movement and other sections with no movement, where the fault is "locked." Using laser beams and an orbiting satellite, scientists at NASA's Goddard Space Flight center have discovered that the rate of ground movement along the San Andreas fault has increased to an average 7.6 centimeters a year. Observations carried out between 1941 and 1975 in the Imperial Valley of southeastern California found that points on the western side of the fault had moved as much as 3.8 meters north of points on the eastern side of the fault. Averaged out, this shift would produce a movement of 11 centimeters a year. These two studies indicate accelerations of between 50 and 100% in the rate of movement along sections of the San Andreas. Yet two areas along the fault seem to be "locked" without any sign of movement. Associated with the sections of the fault that snapped during the 1857 Fort Tejon and 1906 San Francisco earthquakes, these two locked areas could spell catastrophe for San Francisco and Los Angeles when the built-up strain is finally released.

Another geological phenomenon currently occurring along the San Andreas is the "Palmdale Bulge." Located about 60 kilometers north of Los Angeles, a section of the San Andreas fault has been rising since 1961. As if the ground had suddenly popped, the uplift between 1961 and 1962 was 0.2 meters. Since 1961, an area of 83,000 square kilometers has risen close to 0.3 meters. Some scientists fear that this bulge warns of an intense earthquake along this section of the fault. There are other signals, as well.

As discussed earlier in this chapter, John Gribbin and Stephen Plagemann postulate that major earthquakes will be triggered by the grand planetary alignment between 1982 and 1984. This once-every-179-year occurrence, combined with the peaking of the sunspot cycle, leads Gribbin and Plagemann to believe that there will be major earthquake activity throughout the world. They postulate that Los Angeles is liable to experience a devastating earthquake during the early 1980's. In their analysis of eight major earthquakes that hit the San Francisco area since 1836, Gribbin and Plagemann found that each quake occurred within two years after the sunspot cycle had peaked. Although some scientists have greeted this theory of planetary alignment as a catalyst for earthquake activity with skepticism or ridicule, Gribbin and Plagemann are not alone in their prediction of a major earthquake along the San Andreas fault in the near future.

Based on time frequency alone, the two most severe earthquakes in California struck almost 50 years apart—Fort Tejon in 1857 and San Francisco in 1906. As both of these sections of the San Andreas fault are presently locked, and therefore building up unrelieved strain, major earthquakes are likely to hit these areas when the stress finally reaches its breaking point. As days, months, years go by without the release of this accumulated strain, scientists believe the next major earthquake along the San Andreas could be even more severe than the earthquakes of 1857 or 1906. Charles Richter, the dean of American seismology, foresees Los Angeles hit by an earthquake of a magnitude as powerful as 8.5 in the near future. From his study of seismic activity and movement along the San Andreas fault, Robert Wallace of the U.S. Geological Survey predicts that at least one earthquake of 6.0 magnitude will take place along the fault every five years, an earthquake of 7.0 magnitude every 15 years, and an earthquake of 8.0 magnitude once every 100 years. Louis Pakiser, Jr., also of the U.S. Geological Survey, believes that a minimum of 10 damaging quakes will occur along the San Andreas fault in the next 20 years. Although the expected major earthquake along the San Andreas fault has not yet occurred, we should not assume that a severe quake will not strike. It will, and the effects may be devastating. Estimates of casualties from a major earthquake in the vicinity of San Francisco or Los Angeles run to tens of thousands dead and hundreds of thousands injured.

In our consideration of earthquakes and their aftereffects we have focussed primarily on the severe consequences to human life conditions. We have intentionally ignored property damage, for even though an earthquake can cause millions or even billions of dollars worth of damage, the survivors can rebuild. It is difficult to rebuild broken lives, but impossible to bring back the dead. There can be no denial that survivors of a catastrophic earthquake tend to live subsequent lives of fear and anguish. Accounts from survivors of the 1977 Rumanian earthquake tell of the psychological impact from the quakes as far worse than the experiences of the Second World War.

Observations of earthquake victims reveal their tendency to become neurotic. Recurring panic overwhelms them. Agitative depression is common. Survivors feel a sense of futility about their lives and a hopelessness about the future. The will to live may be lost to grief, apathy, and indifference. The civilized manner of society is often replaced by more aggressive, primitive reactions. At an extreme, the social order breaks down. Lawlessness becomes rampant. Life degenerates into a struggle for survival solely for its own sake. The more we are able to learn about earthquake activity, the better prepared we'll be to avoid their ravaging effects. It's a matter of earth alert.

VIII

EARTH ALERT:
PREPARING FOR EARTHQUAKES

R: I can't cut into this; it's as hard as a rock.

L: What sort of a rock?

R: What does it matter? All rocks are the same.

L: But not all rocks are equally hard.

R: You're kidding. What is it that makes the difference?

L: Density—and water pressure. Rocks respond to water pressure by swelling, and this decreases their strength.

R: O.K. Now you've given me a little piece of arcane information, but for all intents and purposes, a rock is still a rock. What use is this information to me? Will it make me more able to cut into this board?

L: No. But it can allow you to detect and prevent earthquakes.

R: Are you trying to tell me that the water pressure in a rock can lead to an earthquake?

L: Not exactly. I'm trying to tell you that the water pressure in a rock can be indicative of the stress in a wider area, and that this stress . . . Well, let me explain by this parable which Italo Calvino relates:

> Marco Polo describes a bridge, stone by stone.
> "But which is the stone which supports the bridge?", asks Kubla Kahn.
> "The bridge is not supported by one stone or another," answers Marco, "but by the line of the arch that they form."
> Kubla Khan remains silent, reflecting. Then he adds: "Why do you speak to me of stones? It is only the arch that matters to me."
> Polo answers: "Without stones there is no arch."

R: I think I understand what you're getting at: the bridge is the earthquake. But you can't stop here with a parable. I want to know about the entire arch of warning signs.

If people could be warned, if they could prepare for the consequences of an earthquake, more lives might be saved. Preparation before the fact would provide an orderly means by which survivors might rebuild their lives. In addition, the psychological impact of unexpected forces threatening our very existence would be lessened. Prediction of earthquake activity is the answer. But such a solution still eludes us at the present. However, our knowledge of the indicators of earthquake activity and their interrelatedness is growing. While predictions of earthquake activity during this advent period have occasionally been wrong, accurate predictions in the recent past have saved tens of thousands of lives and modified the psychological shock on people who have suffered through a major earthquake.

Four countries are actively engaged in major research programs of earthquake prediction techniques. The first national research program started in Japan in the early 1960's, was followed by a national program in the U.S.S.R. and by scattered research in the United States. After the devastating Hsingtai earthquake in 1966, the Chinese began intensive investigations that brought together both professional scientists and amateur observers to record potential precursory events. Among the phenomena believed to be precursors of earthquakes are anomalous changes in the following (1) ground level or tilt, (2) frequency of small earthquake activity or foreshocks, (3) electrical conductivity of rock, (4) local magnetic fields, (5) animal behavior, (6) fluid pressures, (7) well water levels, (8) radon content in well water, and (9) seismic wave velocities. The geophysical precursors all seem linked to a phenomenon known as dilatancy.

Evidence of intense rock strain liable to impending break-age and resulting earthquake can often be detectedby measuring the vertical and horizontal ground movements that change the relative positions of points on the earth's surface. The Japanese have found that before the 6.0 magnitude earthquake at Odaigahara in 1960, instruments began to record increased rates of strain and a sharp tilting of the ground away from the subsequent epicenter. These conditions continued for three months and then stopped three months before the quake. From 1889 to 1955, repeated measurements of the epicentral region of the 1964 Niigata earthquake showed a steady but small rate of uplift. In 1955, this region experienced a sudden uplift of five centimeters. Following this sudden uplift, little movement was recorded for this area until the 7.5 magnitude earthquake struck Niigata on June 16, 1964. Ground movement measurements also forewarned of the earthquake swarm that occurred in the Matsushiro area between 1965 and 1967.

Prior to about a dozen California earthquakes, the epicentral regions experienced ground tilt changes. In their study of the two earthquakes that struck near Danville, California, on June 11 and 12, 1970, Darroll Wood and Rex Allen, both of the U.S. Geological Survey and Stephen Levine, then a graduate student at Stanford University, discovered that ground tilt in the San Francisco Bay area occurred before the quake. Almost a month before the quake, the direction of tilt shifted towards Danville. The degree of tilt toward the subsequent epicenter continued to increase until 10 hours before the quake, at which time the ground movement stopped. After the quake, the direction of ground tilt in the Danville area shifted in the opposite direction. While the total tilt of the region was minimal, it did prove to be a precursor of earthquake activity in the region. Perhaps the "Palmdale Bulge" north of Los Angeles will also prove to be an omen of a significant earthquake in the Los Angeles area. Based on the preceding studies, it becomes clear that regions subject to earthquake activity sometimes undergo vertical or horizontal ground movements beforehand.

To detect vertical and horizontal ground movements, scientists have employed strain meters, tilt meters, repeated ground level surveying, and tidal gauge instruments. While these methods do provide useful information, they suffer from two major deficiencies. Since these instruments primarily measure surface activity, data may be distorted by such variables as meteorological conditions and movements occurring at depths beyond their measurement range. Fortunately, recent technology has found a solution to this: use of lasers and quasars. By bouncing laser beams between two ground stations, small shifts in ground movement can be measured. Developed by North American Aviation's Space and Information Systems Division, Project GLASS (Geodetic Laser Survey System) can measure ground movements as small as 1.016 millimeters. The use of quasars to measure ground movement has been developed under the name of the ARIES technique by scientists at the California Institute of Technology. Due to their great distance from the earth, quasars provide fixed reference points. Two radio antennas situated at different ground locations are used to record the arrival time of radio signals from the quasars. By comparing the arrival times of these radio signals at the two ground locations, the distances between them can be computed. Repeated recording allows for continual measurement, and changes in the arrival times thereby indicate any shift in ground movement. The accuracy of recording is one ten-billionth of a second, and the anticipated goal of measuring ground movement is an accuracy of within 2.5 centimeters. The refinement of this technology will provide us with indications of subtle shifts in ground movement, which in turn will serve as a warning of potential earthquake activity.

In earthquakes of greater than 5.0 magnitude, scientists have found a tendency for the main shock of the quake to be preceded by slight tremors. Following a period of general quiescence, minor seismic activity increases in frequency, but then decreases slightly just before the main shock. If an area begins to register microearthquake activity, it may signal the likelihood of a significant earthquake for that region in the immediate future.

133

Another precursor of earthquake activity is the decreased electrical resistivity of the ground. This phenomenon was the basis for the successful prediction of a major earthquake in Kamchatka in the Soviet Union. In their studies of earthquakes in the Garm region of the Tadzhik Republic of the U.S.S.R., Soviet scientists have found a strong correlation between the minimal electrical resistivity of the ground and the occurrence of earthquakes. Laboratory experiments on rock fracturing have confirmed these findings: fluid-saturated rocks show a strong decrease in electrical resistivity. As water is a better electrical conductor than rock, it is evident that prior to earthquake activity, the rock under stress becomes saturated with fluid. Earthquakes occur when maximal saturation and minimal electrical resistivity of the area are reached.

Scientists have noted that anomalous changes in the local magnetic field sometimes occur before earthquakes, and seem to indicate movements within the earth. An hour before the 1964 Alaska earthquake, magnetometers recorded magnetic field changes. Since disturbances in the local magnetic field are common in Alaska, we might discount this phenomenon as a precursor of earthquake activity, for not all local magnetic field variations have resulted in earthquakes. Nor have all earthquakes been preceded by detected magnetic field disturbances. Nonetheless, evidence does show a correlation between local magnetic field changes and earthquake activity. In 1965 and 1966, magnetometers situated along the San Andreas fault recorded abrupt, radical changes in the local magnetic field prior to five different earthquakes. On November 27, 1974, Malcolm Johnson of the U.S. Geological Survey informally announced his belief that Hollister, California, might experience a quake in the range of about 5.0 magnitude at any time in the immediate future. Johnson made his casual prediction on the basis of the changes in the strength of the local magnetic field and the slight tilt of the surrounding region. One day later, on November 28th, Hollister was hit by a 5.2 magnitude earthquake. Variations in the local magnetic field are insufficient predictions of earthquakes, but together with other anomalous changes in geophysical conditions, they may contribute to a grand scheme of accurate prediction.

Associated with magnetic field changes prior to earthquakes are accounts of abnormal animal behavior starting hours, days, and even weeks before an earthquake. Reports from various countries tell of chickens refusing to roost, cattle panicking in their barns, dogs barking wildly, and rats running dazed into the streets before an earthquake struck. In 1964, before the Alaska quake, Kodiak bears emerged from hibernation two weeks ahead of schedule and moved into the hills. Before a small earthquake in the Stanford, California, area, researchers at the Stanford Outdoor Primate Facility, observed increased excitability and restlessness in the chimpanzees they were studying. Russian scientists have noticed ants picking up their eggs and moving in mass migration, and shrimps crawling onto dry land before earthquake activity. Two hours before the 7.4 magnitude Pohai earthquake in China on July 18, 1969, the zookeeper at Tientsin zoo warned seismologists of a possible imminent earthquake because of abnormal behavior among the animals. The zookeeper had seen the tigers stop their pacing and begin acting strangely, a panda hold its head and moan, a Tibetan yak collapse, and the swans suddenly leave the water. The Japanese have found that before some earthquakes catfish, which dwell at river bottoms, leap out of the water or move into unusual river habitats.

How can animals sense impending earthquake activity? Through their sensitivity to magnetic field changes. As noted in Chapter VIII, James D. Hays has found a susceptibility to variations in the magnetic field in such species as fruitflies, mud snails, and flatworms. Researchers have also discovered magnetic field sensitivity among honeybees, European robins, seagulls, and pigeons. While directional orientation may be thrown off by changes in the magnetic field, it seems as if these species are able to correct their orientation. Not only does abnormal animal behavior give credence to local magnetic field changes before earthquakes, but it also provides another signal of impending earthquake activity.

Strange changes in subterranean fluid pressures may also forewarn of imminent earthquake activity. Before certain earthquakes, scientists have witnessed unexpected variations in the flow of springs and streams. Wells have also shown radical fluctuations in their fluid level. In their study of the Izu-Hanto-oki earthquake of May 9, 1974, Japanese scientists noted prior changes in the water level in 59 out of the 95 observation wells located in the districts of Kanto and Tokai. Anomalous changes in fluid pressures and levels may indicate undetected alterations in the strain on subterranean rock. The radon content in well water gives further credence to accumulating stresses within the earth's crust.

Radon is a radioactive gas, released by the decay of radioactive minerals in rock. It dissolves in the waters of deep wells. When rocks are under intense stress, they release greater amounts of radon. Since radon has a half-life of only 3.8 days, and its distance of diffusion is limited to a few centimeters, mointoring the concentration of radon in well water provides a reliable gauge of significant changes in the pressure of the subterranean rock in a local area. Before the 5.3 magnitude Tashkent earthquake of 1966, the Soviet scientist V.I. Ulomov called attention to the increase of the radon concentration in the mineral waters of the area. Known for their medicinal properties, the chemistry of these waters has been continuously monitored since 1956. By the middle of 1965, Ulomov found that the radon content of these waters had almost doubled from its normal concentration. Through September, 1965, the radon concentration increased sharply and then levelled off until the earthquake struck on April 26, 1966. After the quake, the radon content dropped significantly. The same phenomenon occurred before the 4.0 magnitude aftershock in 1967, when the radon content jumped before the quake and dropped after it. Scientists believe radon concentration in well water may prove a useful tool in predicting both the approximate magnitude and the timing of an earthquake.

A major breakthrough in understanding warning signs of earthquake activity came with the discovery of the anomalous seismic wave velocities that occur before earthquakes. In 1969, two Soviet scientists, A.N. Semenov and I.L. Nersesov, both of the Institute of Physics of the Earth in Moscow, found that unusual changes in the ratio between the compression wave velocities (V_p) and the shear wave velocities (V_s) occurred before several moderate earthquakes in the Garm region of the Tadzhik Republic of the U.S.S.R. Although the V_p: V_s ratio is normally 1.75, it decreased by about 6% during the period preceding the quakes. Just before the quakes struck, the V_p: V_s ration returned to its previous level of about 1.75. In his study of the 1971 earthquake swarm in the Blue Mountain Lake region of the Adirondacks in New York State, Yash

Aggarwal of the Lamont-Doherty Geological Observatory of Columbia University discovered a similar phenomenon occurring before smaller earthquakes. During this earthquake swarm, the V_p: V_s ratio decreased by as much as 13% before earthquake activity, and returned to its normal value just before the quakes hit. The same situation happened before the 1971 San Fernando, California, earthquake, where a 10% decrease in the seismic wave velocities ratio began three and a half years before the quake. From their analysis of these observations, James Whitcomb, Jan Garmany, and Don Anderson, all of the California Institute of Technology, believe that the duration between the initial decrease and the actual earthquake provides accurate information on the magnitude of the earthquake. The longer the duration, the greater the magnitude. When calculated, it seems that the warning time is about 30 days for earthquakes of 5.5 magnitude, about 3½ years for a 6.5 magnitude earthquake, and about 10 years for a quake of 7.5 magnitude. Not only did this discovery provide another important tool for estimating warning signs, but it offered convincing support to the process of dilatancy, a phenomenon long noted in laboratory experiments on rock fracturing.

First discovered in 1886 by Osborne Reynolds, a British mechanical engineer at the University of Manchester, and studied in great detail in the mid-1960's by William Brace and his colleagues in their laboratory experiments on rock fracturing, dilatancy is a phenomenon whereby rock under pressure tends to swell, eventually creating cracks and pores. As that total crack volume of the rock increases, there is more space for fluids within it. Increased volume lessens the pressure of fluids on the rock. Decreased pore pressure strengthens the rock. When fluids from the surrounding area fill the increased volume of pores and cracks, the fluid pressure of the rock returns to its normal value. The rock, initially strengthened by swelling, is now weakened, and earthquakes are likely to be triggered. Based upon Brace's experiments, and findings by Semenov and Nersesov described above, several American scientists have proposed various models of dilatancy as the process which precedes earthquakes.

Anomalous changes in geophysical phenomena are linked to dilatancy, due to the swelling of the rocks. Bulging is seen in the ground tilt or vertical movement of the subsequent epicentral region. The increase of minor seismic activity, followed by a slight decrease before the main shock of the quake, adheres to the dilatancy pattern of accumulated stress. There is then a decrease of fluid pressure from the volume increase of the rock, and finally saturation, which returns the pore pressure to its normal value and triggers the earthquake. When dilatancy begins, the increased pores and cracks create more air within the rock and the electrical resistance of the rock rises. Conversely, as the increased volume of the rock becomes saturated, the electrical resistance decreases. Since water is a better conductor of electricity than rock, electrical resistance will be least when the saturation of the rock reaches its maximum, just before the rock breaks and triggers the earthquake. Subterranean fluid pressures will fluctuate, because the increased porosity of the rock will reduce the fluid pressure until the expanded volume of the rock is saturated, at which time the fluid pressure will return to normal. Water levels will change due to the inflow of adjacent waters into the rock. The radon content of well water increases during dilatancy, because the water is exposed to more of the radioactive minerals within the stressed rock. As the rock reaches its maximal saturation, the radon concentration in the well water will level off. The anomalous seismic wave velocities occur due to the undersaturation of the swelling rock during the initial stages of dilatancy. While this condition does not affect the velocity of the shear waves, it does reduce the velocity of the compression waves. Consequently, the $V_p: V_s$ ratio of 1.75 drops. After the swollen rock is saturated, the velocity of the compression waves rises back to its usual speed, and the $V_p: V_s$ ratio returns to its normal value, just before the earthquake takes place.

The theory of dilatancy and its geophysical effects provide an important clue to impending earthquake activity. Scientists believe they can gauge the magnitude and the approximate timing of the subsequent earthquake from the breadth and time duration of the dilatancy effects. The larger the region over which these effects take place, the greater the quake and the longer it will take to occur. Some estimates calculate that the quake will occur within about one-tenth the time of these various anomalous changes are recorded. For example, if changes in geophysical phenomena were measured over a period of 30 days before they returned to their normal values, it is likely the quake would occur three days after the return to normal values. If changes were measured over 90 days before returning to normal, the quake is likely to occur nine days after these phenomena returned to normal. The latter earthquake would also be expected to be of greater magnitude than the former. However, earthquakes will not always occur after recording such changes. The accumulated strain could be released gradually through slow creep of the fault, instead of through the violent release of an earthquake. If lives can be saved and human suffering mitigated by a warning system based upon this system of prediction, then the anxiety and cost of preparation for one false alarm or two incorrect predictions would seem worth the price for one correct prediction. As our understanding of our mother earth and her dynamics increases, we shall continue to refine our techniques of prediction for greater accuracy. The recent past has already witnessed successful prediction of earthquakes. The most successful, in terms of timing accuracy and the number of lives saved as a result, was the Chinese prediction of the 1975 earthquake in the Haicheng-Yingkou region of Liaoning Province in China.

After the severe earthquake of 1966, which struck Hsingtai in Hopei Province, about 300 kilometers southwest of Peking, the Chinese placed a high priority on earthquake prediction. Their research program enlisted the efforts of 10,000 professional seismologists and 100,000 amateurs to serve as observers in the field. To record and measure any change in geophysical phenomena relevant to earthquake activity, the Chinese established a network of 17 major seismograph stations, 300 auxiliary regional stations, and 5000 observation points.

Following the Hsingtai earthquake, epicenters of subsequent smaller earthquakes were observed to be moving in a northeasterly direction toward the heavily industrial and highly populated Liaoning Province. In 1970, the Chinese decided to maintain a close watch on the province for any signs of earthquake activity. Temporary networks of seismic instruments were installed to monitor geophysical activities. Between 1970 and 1974, studies of seismic activity, geophysical investigations into the local structure of the earth's interior, geological fieldwork to map faults, and continual ground level surveys were carried out. From the records of earthquake activity dating back 3000 years, it became apparent that the frequency of earthquake activity in the region was sharply increasing over the norm. Several faults in the province were found to be active, and the ground level was tilting, with the southeastern area of Liaotung Peninsula rising in relation to the northwestern segment near Yingkou. Changes were also recorded in the sea level and in the local magnetic field intensity.

During 1974, the warnings of earthquake activity became more noticeable. Ground tilt in a northwesterly direction was rising twenty times faster than normal. In 1974, there were five times as many earthquakes as usual. It became increasingly evident that the region would be struck by a major earthquake. In June, 1974, the State Seismological Bureau issued a prediction for an earthquake of between 5.0 and 6.0 magnitude in the region within one or two years. At this stage, the Chinese political leadership began a massive program to educate the public about the nature of earthquakes, the techniques of earthquake prediction, and the measures to be taken to reduce the calamitous effects of an earthquake. The people became involved in monitoring and reporting any anomalous changes that signaled imminent activity.

By the end of 1974, ground deformation was accelerating. Water level changes in wells were noted, with the water becoming muddier and starting to bubble. The radon content in ground water increased by 20 to 40%. Strange behavior was observed among the animals in the region. The previously increasing seismic activity decreased sharply. On December 22, a 4.8 magnitude earthquake struck 70 kilometers north of Haicheng. This earthquake was presumed to be the prelude to an even larger quake.

These precursory earthquake phenomena increased, and in mid-January 1975, the Seismological Bureau issued another prediction. During the first half of 1975, the bureau warned that Liaoning Province, and specifically the Liaotung Peninsula within the area bordered by the cities of Dairen, Yingkou, and Antung, would likely experience an earthquake of between 5.5 and 6.0 magnitude. Further preparations for an earthquake were ordered. Contingency plans to mitigate the potential of disaster were drawn up. Disaster relief and emergency medical care plans were readied. Authorities inspected local buildings, mines, and reservoirs for structural weakness, and repaired the deficiencies.

At the beginning of February, it became evident that a major earthquake was close at hand. The anomalous changes in precursory phenomena intensified. Ground tilt shifted direction, well water levels fluctuated wildly, electrical resistivity dropped sharply, local animals displayed weird behavior. On February 1, a 0.5 magnitude tremor was detected. Small earthquake tremors, recorded in an area of previous seismic activity, were believed to be foreshocks of a large quake. Between the evening of February 3 and midday on February 4, 466 foreshocks were recorded. With two tremors registering in the range of 4.0 to 5.0 magnitude, the morning of February 4 witnessed a peaking of seismic activity, and a radical decrease thereafter. The authorities interpreted this quiescence as the calm before the storm.

At 10:00 A.M. on February 4, the Provincial Revolutionary Committee declared an earthquake emergency. The contingency plans devised in January were put into operation. An evacuation order went out. The citizens began to build huts and to erect tents to serve as temporary shelters. The old and infirm were assisted in their evacuation. Motor vehicles were driven from garages and grouped in the open. Livestock were removed from barns and placed in open-air corrals. In some communes, movies were shown in open areas for the populace. At 7:36 P.M., an earthquake of 7.3 magnitude struck the Haicheng-Yingkou region. Although damage to building structures was severe, and virtually total in some areas, the mass evacuation of over one million people minimized the death toll. Two to three hundred people were killed in this earthquake. But its prediction, and the precautions taken, are estimated to have saved the lives of tens of thousands.

The Chinese have had mixed results in their accuracy of earthquake prediction. False alarms have been issued, and major earthquakes have occurred without warning. While Chinese scientists warned early in 1976 of an impending severe earthquake in the Peking or Tangshan area, conflicting information led to the failure to predict specifically the Tangshan earthquake of July 28, 1976. In terms of human casualties, this earthquake proved to be the second worst in recorded history. Estimates of the death toll run as high as 800,000. Yet, the Chinese are credited with saving many lives by their accurate prediction of two other severe earthquakes with magnitudes

Map of Northern China, showing the earthquakes at Haicheng-Yingkow (February 4, 1975) and Tangshan (July 27, 1976).

registering 7.5 and 7.6 that struck Lungling County in western Yunnan Province on May 29, 1976. There is no denial that earthquake prediction is still a young science. However, our knowledge is increasing daily, and it would seem that saving human lives is worth the cost of an error or two. Unfortunately, some do not agree.

In a paper presented on April 15, 1976, James Whitcomb, a highly respected seismologist, noted the possibility of an earthquake with a magnitude between 5.5 and 6.5 striking the Los Angeles area within the year. An immediate consequence of Whitcomb's announcement was the threat of a lawsuit by a Los Angeles city councilman, who feared the economic impact of such a prediction. One cannot deny the economic impact on a region threatened by major earthquake. Land values are likely to plummet. The local economy may be harshly affected. The social fiber of the community might break down from the anxiety of anticipation. Nonetheless, the adverse economic impact should be outweighed by the opportunity of saving lives and lessening the effects of the disaster. Politics and economics must yield to the preparation for eventual earthquakes. Nature is not solely responsible for the upheavals that occur.

By his own actions, man has inadvertently triggered earthquake activity. During the 10 years following the damming of the Colorado River and the forming of Lake Mead by the construction of Hoover Dam on the Arizona-Nevada state line, approximately 600 earthquake tremors were documented in the area where no earthquake activity had been recorded during the 15 years before construction. In his study of the correlation between the construction of the Hoover Dam and local earthquake activity, D.S. Carder of the U.S. Coast and Geodetic Survey concluded in 1945 that the increased load of water in Lake Mead had reactivated faults in the area and thereby triggered the earth tremors. While most of these quakes were small, two of them registered approximately 4.0 magnitude, and a third was of 5.0 magnitude. Since Carder's findings, similar examples have appeared.

On the border of Zimbabwe and Zambia in the Kariba Gorge, the Zambezi River was blocked by the construction of a hydroelectric dam in 1958. As Lake Kariba began to fill behind the 128-meter high dam, this previously aseismic area began to experience tremors. Through September 1963, when the reservoir finished filling and the seismic activity peaked, over 2000 quakes were noted in the area. In the last week of September 1963, five major shocks, registering between 5.6 and 6.1 magnitudes, were recorded in the vicinity of the dam. Although these quakes and those associated with Hoover Dam in the U.S. did not cause any deaths, an earthquake linked to the construction of Koyna Dam in India did result in close to 200 deaths. Koyna Dam was constructed in 1962 in an area once described as one of the least seismic in the world. As the dam began to fill, seismic activity occurred with increasing frequency. On December 10, 1967, an earthquake of 6.4 magnitude struck the region, killing at least 177 people and injuring close to 2000.

The cyclical frequency of the Koyna quake activity lends support to the notion that the water load triggers the tremors. Each year the seismic activity increases after the rainy season, when the reservoir reaches its highest level. There are other examples which corroborate this. Greece is prone to earthquakes, but the construction of dams at Kremasta and Marathon was followed by increases in seismic activity corresponding to the water level of the lakes. In 1959, the Hsingfengkiang Dam was completed north of Canton in China; shortly thereafter the region around the dam began to experience an increasing frequency of minor earthquakes. In 1972, more than 250,000 small tremors were registered in this vicinity. As soon as water began to fill behind the 317-meter high Nurek Dam in Tadzhikistan in the U.S.S.R., local seismic activity increased. Oued Fodda in Algeria and Monteyard Dam in the French Alps tell a similar story.

Dams and reservoirs have been shown to generate earthquakes. The evidence indicates that the higher the dam the greater the incidence of activity. This correlation between dams and earthquakes can be linked to the concept of dilatancy. The construction of a dam increases the load on the underlying rock in the area. This condition intensifies any accumulated strain within the rock, which causes the rock to swell, creating new cracks and pores. The increased total crack volume will initially cause the fluid pressure of the rock to drop until seepage from the dammed water brings the fluid pressure back to its normal value. The rock will then be weakened, and earthquake activity may be triggered.

Underground nuclear explosions are another means by which man inadvertently triggers earthquake activity. When a nuclear device is exploded, it releases an incredibly large amount of energy, causing quantum jumps in the pressure and the temperature in the surrounding rock. The explosion vaporizes the surrounding rock and creates a cavity. Further out from this cavity the compression of the rock produces seismic waves. Although most quakes associated with nuclear explosions are of smaller magnitude than the explosion itself, the amount of strain released depends both on the type of rock and the accumulated strain existing in the area. These earthquakes extend only about 20 kilometers from the shot point. However, one cannot rule out the possibility of a major earthquake being generated by a nuclear explosion in an area of intense strain.

Such a catastrophe has not yet occurred, but sometimes it seems as if man is purposely attempting the feat.

Our studies of the correlation between underground nuclear explosions and earthquake activity are based primarily on American tests conducted at the Nevada Test Site and on Amchitka Island in the Aleutians. The United States has encouraged scientific study of its tests to learn the geophysical ramifications of underground nuclear explosions. In 1968, the Atomic Energy Commission set off two well-studied nuclear devices at its Nevada Test Site. On April 26, 1968, the ''Boxcar'' nuclear test detonated a 1.2 megaton explosion. The shaking from this explosion was felt as far away as Las Vegas, at a distance of 50 kilometers. Within three days after detonation, 30 small squakes were registered around the test site, and during the following month, thousands of minor tremors were recorded. Alan Ryall of the University of Nevada studied this explosion, and showed that the aftershocks were concentrated in a zone 12 kilometers long and four kilometers wide, suggesting the reactivation of a fault in the area. On December 19, 1968, the ''Benham'' nuclear test detonated a 1.1 megaton explosion, which caused movement along surrounding faults and generated earthquake activity that lasted several months. During the four weeks following Benham, approximately 10,000 aftershocks were recorded as far away as 13 kilometers from the shot point, and at focal depths varying from ground surface to seven kilometers.

The United States has also tested nuclear devices on Amchitka in the Aleutian Islands. Although geological studies of Amchitka have found it to be free of natural earthquake activity, the Aleutian Islands form part of the Pacific Ring of Fire, and consequently are a prime area for seismic activity. When the United States announced its intention to test a series of nuclear devices at Amchitka, the governments of Canada and Japan officially protested, and were joined by a number of scientists who feared that nuclear testing on Amchitka was likely to trigger severe earthquakes in the region. In October 1969, the United States went ahead with its project and detonated the ''Milrow'' nuclear test there, a 1.2 megaton explosion. Immediately afterward, hundreds of minor shallow-focus quakes occurred within an area of five kilometers from the shot point. The aftershock activity ended suddenly, upon collapse of the cavity produced by the explosion. While the blast and its cavity collapse were equivalent to 6.5 and 4.3 magnitudes, respectively, only two of the aftershocks were larger than 3.0 magnitude, the largest registering about 3.4.

In November 1971, the United States detonated the "Cannikin" nuclear test on Amchitka. The decision to test nuclear devices there was once again protested by scientists, but to no avail. As before, a swarm of hundreds of aftershocks occurred, and were associated with the cavity collapse from the explosion. But 22 "natural" earthquakes resulted from the readjustments of the strain within the earth's crust. The largest of these subsequent quakes, registering 3.5 magnitude, struck more than seven days after detonation. Therefore, we should not neglect the possibility that nuclear explosions can cause significant underground structural changes that may initially go undetected but may eventually trigger a catastrophic earthquake. So far, man has been lucky, but his luck could run out. Even more fearsome is the recognition among the scientific community that strategically planned explosions by an antagonistic country could trigger a major earthquake in a neighboring country, or, farther down, a seismically liable zone.

Man has also set off earthquakes by pumping fluids into underground reservoirs. The correlation between such fluid injection and earthquake activity was shown by the experience at the Rocky Mountain Arsenal near Denver, Colorado. In 1961, the U.S. army drilled a well to a depth of 3.67 kilometers at the Rocky Mountain Arsenal to dispose of chemical waste fluids. From March 8, 1962, to September 30, 1963, waste fluids were pumped into this underground reservoir. Although the region around Denver had not experienced any detectable earthquake activity since 1882, local seismographs between April 1962 and September 1963 recorded 710 earthquakes, with magnitudes ranging between 0.7 and 4.3 on the Richter scale in the vicinity of the Rocky Mountain Arsenal. Between September 1963 and September 1964, injection of fluids into the reservoir ceased. During this period, there was a steep decline in earthquake activity. In September 1964, pumping resumed. Again, a series of quakes was recorded. Even though pumping was abandoned in February 1966, with a consequent initial reduction in earthquakes, activity in the area was recorded during the last part of 1966 and through most of 1967 before declining in 1968. In 1967, earthquakes with magnitudes of about 5.0

struck the area on April 10, August 9, and November 26. In his study of these phenomena, David Evans, a consulting geologist, demonstrated the correlation between earthquake activity and fluid injection into underground reservoirs. To explain this correlation, Evans cited the theory of M. King Hubbert of Stanford University and William Rubey of the University of California at Los Angeles that increased fluid pressure on the pores of deep rock will facilitate strain release in the form of earthquake activity.

To test the correlation between the pumping of fluids underground and earthquake activity, scientists with the U.S. Geological Survey performed planned experiments at the Rangely Oil Fields in Colorado, where the Chevron Oil Company had been engaged in secondary recovery operations, forcing oil to the surface by the injection of fluids. As at the Rocky Mountain Arsenal, this operation seemed to be generating earthquake activity in the area. The strongest of these quakes occurred near wells subjected to the greatest injected fluid pressure. From September 1969 through May 1973, government scientists took over four wells at the site for their experiments. From October 1969 to November 1970, the researchers injected fluid into the wells. More than 900 earthquakes were recorded at the site, more than a third of them within a kilometer of the bottom of the injection wells. In November 1970, injection was stopped, and the researchers began removing water from the ground by means of backflow. This action reduced the fluid pressure on the rocks, and consequently the seismic activity in the field dropped. Whereas the average seismic activity within a kilometer of the experimental wells had been 28 quakes a month for the preceding year, such seismic activity decreased to about one quake a month. When water was reinjected into the wells between June 1971 and April 1973, seismic activity again increased. After fluid injection was halted, no earthquake activity was reported in the area. The experiments at Rangely not only confirmed the correlation between fluid injection underground and earthquake activity, but also led scientists to believe they had found a means to control earthquakes.

Through the concept of dilatancy and the evidence of fluid injection, we know that rock is strengthened by decreasing fluid pressure and weakened by increasing fluid pressure. Scientists have proposed using this knowledge to relieve accumulated strain in the earth's crust before it is suddenly released in a violent earthquake. Barry Raleigh of the U.S. Geological Survey, has suggested boring three holes about half a kilometer apart at depths of about four kilometers along a potentially dangerous fault. By pumping water out of the outer holes, the fault would be locked at those points. Water would then be injected into the middle hole, increasing the fluid pressure in that area and triggering earthquake activity. The quakes triggered, however, would theoretically be contained within the parameters of the two outer holes. Once the strain was relieved through the minor earthquakes generated, water could be pumped out of the middle hole to lock the fault in that area more securely. Another proposal for earthquake control suggests the use of small underground nuclear explosions to trigger minor quakes and thereby release the accumulating strain along a fault before it builds to the point of a major earthquake. In the abstract, these two concepts seem to provide the answer to control of earthquake activity, but they might have unknown inherent flaws when put into practice. Meddling with the forces of nature is at least a dangerous risk, if not a prelude to catastrophe.

Perhaps the best solutions for mitigating the effects of earthquakes are an awareness of the consequences of our own actions and orderly preparation for possible earthquake activity in the future. From past experience, we know that the construction of dams and reservoirs can produce increased load factor and water seepage that generate earthquake activity. We know that underground nuclear explosions have set off series of aftershocks, due both to cavity collapse and structural readjustments of the earth's crust. We know that the injection of fluids underground acts as a catalyst of earthquake activity. Indeed, we know a good deal from past experiences. But will we learn from them? In the future, we might consider the geological history of an area before we construct dams. We could question the needs versus the risks before detonating nuclear explosions. Before secondary recovery operations to squeeze the last amount of oil from a well, we should analyze the region for the potential damage from earthquake activity caused by fluid injection. Whether man will ask the necessary questions, and insert the significant variables, could decide his future. A case in point is the question of the use of nuclear energy and the construction of nuclear power plants.

Few would deny the inherent hazards presently associated with the use of nuclear energy. The proliferation of nuclear weapons, the disposal of nuclear waste, and the danger from inadvertent leakage of radioactive materials are thorny questions whose perfect answers may lie beyond our ken. Despite this, man has gone ahead and constructed nuclear power plants. In deciding on the locations for nuclear power plants, man seems to go blindly into the night. The placement of the nuclear plants at Indian Point, New York, provides a good example of man's readiness to gamble with his survival. These three nuclear units are situated within one kilometer of a major branch of the Ramapo fault system, which government authorities assume is inactive. In its "Final Environmental Statement Related to Operation of Indian Point Nuclear Generating Plant Unit No. 2," the U.S. Atomic Energy Commission in 1972 stated: "There are no truly major faults in or near the site."

Although it would provide some sense of relief if this statement were true, evidence suggests otherwise. Not only does the Ramapo fault system run near the site, but in 1975 Dr. Nicholas Ratcliffe presented evidence that a fault, possibly a branch of the Ramapo system, lies directly beneath reactor unit number 3. Since 1976, microearthquake tremors have been recorded both to the southwest and northeast of the Indian Point site, and even beneath the site itself.

In their study of the Ramapo fault system and the dangers it poses to the Indian Point units, Yash Aggarwal and Lynn Sykes, both of Columbia University, remind us that past severe earthquakes have struck along this fault. Using the Modified Mercalli scale as the basis of measurement, Aggarwal and Sykes suggest the Ramapo system to be the fault along which several of the six large earthquakes around the New York City area have taken place. The quakes of 1737, 1884, and 1927 are listed as having Modified Mercalli (MM) intensity of VII, while the quakes of 1783, 1895, and 1957 are given MM intensities of VI. Although the Indian Point plants are designed to withstand an earthquake of MM intensity VII, Aggarwal and Sykes believe that over their projected 40-year lifetime the area around them stands a 5 to 11% probability of experiencing an earthquake equal to or greater than MM intensity VII. They project a 2% probability for an earthquake equal to or greater than MM intensity VIII. Granted, these probabilities are slight.

149

The tens of thousands or even hundreds of thousands of human lives at stake, however, must not be ignored. Indian Point not the only example.

In New England, nuclear power plants in East Haddam, Connecticut; Plymouth, Massachusetts; and Seabrook, New Hampshire, are also believed to be located near sites of past earthquake activity. If we could be sure that earthquakes will not strike near these locations again, we might be able to feign a sense of security. Unfortunately, scientists think that earthquake activity, like a sleeping giant awakened, is rising out of a quiescent stage and entering a more active period. Not only is the frequency of earthquake activity likely to accelerate in the future, but the magnitudes of earthquakes may increase as well. Man has an unrelenting tendency to remedy perilous situations after it's too late. We must not wait for disaster to strike before we take corrective action.

As a result of the 1933 Long Beach earthquake, which claimed 120 lives, the California state legislature passed the Field Act, requiring newly constructed public school buildings to meet earthquake safety standards. The wisdom of this act was confirmed by the Kern County earthquake of 1952, when damage to school buildings was confined to those constructed before the Field Act. In areas liable to earthquake activity, structures should be designed to withstand shocks of magnitudes greater than those experienced in the past.

The Imperial Hotel in Tokyo, completed in 1922, was designed by Frank Lloyd Wright as a flexible structure. In 1923, the Kwanto earthquake occurred off the coast of Honshu. The quake devastated the cities of Yokohama and Tokyo, but the Imperial Hotel suffered only a few cracks and was turned into a major medical center to care for Tokyo survivors of the quake. By incorporating flexibility for lateral and vertical movements in the design of buildings, architects can miminize potential damage from earthquake activity.

One of the most important aspects in the preparation for earthquake activity is to educate the public. We have noted the Chinese program of massive public education. Other nations, also vulnerable to earthquake activity, must follow their lead. Governments should encourage research into earthquake prediction and be willing to accept long-term human safety instead of short-term economic profit. The impact of an earthquake is lessened if the population of an affected region is prepared. Although readiness for increased earthquake activity might sound like a prescription for a doomsday scenario, it's apparent to scientists that the future is going to demand this.

IX

BREATHING IN AND
BREATHING OUT:
THE EARTH'S
CLIMATE MACHINE

R: Should I take my umbrella?

L: Yes, the weather forecast said there's a 75% chance of rain this afternoon.

R: And since the weather forecast is right about 50% of the time, that means there is 50% of a 75% chance that it will rain. I never pay attention to those forecasts; they're just guesses.

L: You know that's not true. The weather reports are based on data recorded by very sophisticated equipment and analyzed by well-trained meteorologists.

R: Then why are they wrong so often?

L: Because there are so many variables that can reverse the trends and because we've not been studying the weather long enough to be able to . . .

R: We've been recording data for at least 100 years.

L: How impatient you are. A hundred years is a very short period of time in the larger scheme of the earth's changes. Philip Larkin spoke about this impatient attitude of man when he wrote:

> Truly, though our element is time,
> We are not suited to the long perspectives
> Open at each instant of our lives.
> They link us to our losses.

R: What kind of losses is he talking about?

L: Loss of stature, loss of a feeling of self-importance. Man does not like to see himself dwarfed by time—nor space, either, for that matter. Weather is affected by forces as distant as sunspot activity, and you've got to admit that from the sun, man looks very small indeed.

R: But can we continue to ignore this perspective, refusing to look at ourselves from a distance?

L: We certainly can, but if we do so, we'll really lose control of our future.

R: O.K., I'll take my umbrella.

Through the news media and our personal experience, we are repeatedly made aware of the awesome powers of the natural forces of the weather. Winter blizzards paralyze our activities. Summer droughts parch our crops. Floods ravage the countryside, sweeping away our hard-fought prosperity. Weather extremes have wrecked our ambitions and threatened our very survival. Although continually confronted by the violence of the weather, we tend to portray weather extremes as "abnormal," as if there were some sacred mean of normalcy. So-called normal weather describes average conditions for an area over the preceding 30 years. In many areas of the world, weather records date back only 200 years. Be it 30 years or even 200 years, these time periods as a frame of weather reference are but a minute fraction of the earth's climate history.

While computer technology and observation satellites have helped us make great strides in our quest to comprehend climate, the diversity of opinion concerning future trends demonstrates the present groping to understand the various processes that determine our physical environment. Forecasts of future climate conditions range from the extreme of an ice-laden planet to an infernolike earth. Between these two extremes are an assortment of frequently contradictory proposals and opinions. Why such a diversity among the scientific community? The lack of consensus springs both from our limited knowledge of the mechanics of weather and from the question of what importance to place on the variables that influence climate. Let us turn initially to the earth's atmosphere.

Extending from the earth's surface to a height of about 800 kilometers, the atmosphere is a gaseous envelope held in place by gravitational force. From the surface outward, the atmosphere is divided into four basic layers of (1) troposphere, (2) stratosphere, (3) mesosphere, and (4) ionosphere, also known as the thermosphere.

While the interdependence of the four atmospheric layers determines the earth's climate conditions, the troposphere is the atmospheric layer largely responsible for our weather patterns. The upper level of the troposphere ranges from about 10 kilometers near the earth's poles to about 16 kilometers in the tropics. As we will see later, the variance of solar radiation in heating the earth's surface at the equator and at the poles generates the convective process of atmospheric circulation. This atmospheric convection is augmented by temperature differences at varying heights of the troposphere. As the average temperature in the troposphere decreases with height, the denser, colder air at the higher altitudes tends to fall, increasing the turbulence of the convection cycle.

At the tropopause, the boundary between the troposphere and the stratosphere, there is a temperature reversal. In the stratosphere, temperature increases with height. In contrast to the troposphere, the stratosphere is a more stable atmospheric layer. Air movements are slower and far less turbulent here than in the troposphere. An interchange exists between these two atmospheric layers. The circulation of the lower stratosphere is strongly influenced by tropospheric circulation, and sudden warmings in the stratosphere tend to precede warmings in the troposphere by about two days. It's the ozone which is present in the stratosphere that absorbs the higher, ultraviolet

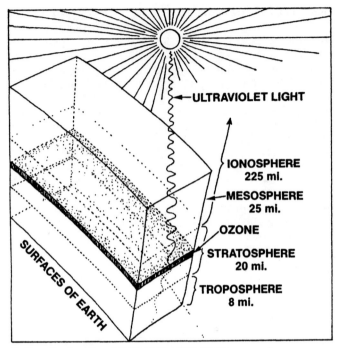

Schematic drawing of the earth's atmosphere.

frequencies of lethal solar radiation, preventing it from reaching the earth's surface. The stratosphere extends to about 50 kilometers above the earth's surface, where it meets the stratopause, the boundary between the stratosphere and mesosphere.

In the mesosphere, another temperature reversal takes place, for here the temperature decreases with height. A relatively quiet atmospheric layer, the mesosphere extends approximately from 50 kilometers to 80 kilometers above the earth's surface where we find the mesopause, the border between the mesosphere and ionosphere.

Temperature reversal again takes place, for in the ionosphere the temperature inceases with height. Extending to about

155

320 kilometers above the earth's surface, the ionosphere derives its energy from intense solar radiation. In this atmospheric layer, the solar radiation strips electrons from nitrogen and oxygen atoms. This process creates positive- and negative-charged particles, thus making the ionosphere a conductor of electricity. It is due to the electrical conductivity of the ionosphere that, by reflecting radio waves off this atmospheric layer, long distance communication was possible before communications satellite technology. The electric currents in the ionosphere also generate magnetic fields. Because of the variability of solar activity and the consequent variability of the ionospheric electric currents, radio communications can at times be disrupted. The ionosphere can produce transient disturbances to the earth's magnetic field, as well. Although the ionosphere is the outer layer of the earth's atmosphere, slight traces of atmosphere extend beyond the ionosphere to a distance of 800 kilometers from the earth's surface in an area called the exosphere.

In the lower atmosphere, the major components of air as given in the proportion of their volume are:

Nitrogen	(N_2)	78%
Oxygen	(O_2)	21%
Argon	(Ar)	0.93%
Carbon Dioxide	(CO_2)	0.03%
Other constituents (see Table 1)		0.04%

TABLE 1
OTHER CONSTITUENTS OF AIR IN THE
LOWER ATMOSPHERE

Water Vapor	(H_2O)
Neon	(Ne)
Helium	(He)
Methane	(CH_4)
Krypton	(Kr)
Hydrogen	(H_2)
Nitrous Oxide	(N_2O)
Carbon Monoxide	(CO)
Xenon	(Xe)
Radon	(Rn)
Ozone	(O_3)
Nitrogen Dioxide	(NO_2)
Nitric Oxide	(NO)
Sulfur Dioxide	(SO_2)
Hydrogen Sulfide	(H_2S)
Ammonia	(NH_3)
Formaldehyde	(CH_2O)
Nitric Acid	(HNO_3)
Methyl Chloride	(CH_3Cl)
Hydrochloric Acid	(HCl)
Freon-11	$(CFCl_3)$
Freon-12	(CF_2Cl_2)
Carbon Tetrachloride	(CCl_4)

Since the sun is the vital center of our solar system, the balance between incoming solar radiation and outgoing terrestrial radiation maintains the earth's climate in a state of relative equilibrium. Solar radiation provides our earth with the warmth that allows life as we know it to exist. The hardest desert on earth is more hospitable than any environment on Mars or Venus. Note, however, that not all the solar radiation directed toward the earth reaches the earth's surface.

Only about 43 to 50% of the solar energy at the edge of the earth's atmosphere reaches the earth's surface. As solar radiation makes its way through the earth's atmosphere, some is absorbed and some scattered by the gases and particles that comprise the atmosphere. The atmospheric absorption of solar energy includes the ultraviolet frequencies of solar radiation, as in the case of the stratospheric ozone that shields the earth from these higher solar frequencies harmful and potentially lethal to earth life. The scattering of solar radiation redirects the rays in all directions. Some of the scattered rays hit other atmospheric gases and particles and are again redirected. While some scattered rays will reach the earth's surface, others will be redirected back toward interstellar space.

At the earth's surface, a similar process occurs. Some of the solar energy is absorbed and some is reflected. The outgoing terrestrial radiation is the reflection of solar energy and the emission of the earth's energy. All bodies radiate energy. The wavelength of radiation emitted is contingent on the temperature of the radiating body. Since the earth is cooler than the sun, the earth's radiation is a lower frequency with a longer wavelength than the solar radiation received. Some of the outgoing terrestrial radiation is absorbed by the atmosphere. This absorption reradiates the outgoing energy both upward, away from the earth's surface, and back, downward toward the earth's surface.

In considering a balance between incoming and outgoing radiation, we are not dealing with a static equilibrium, but rather a process of balancing a myriad of variables. When this balance is affected, it changes the temperature at the earth's surface. For instance, should the solar energy output increase, the mean surface temperature of the earth would also increase. When the incoming solar radiation decreases, the mean surface temperature decreases.

On a local level, we witness seasonal shifts in temperatures due to the inclination of the earth's axis during its yearly orbit around the sun. The distance between a given area on the earth and the sun changes throughout the year, consequently increasing or diminishing the area's solar radiation. This process results in seasonal variations which are more pronounced in the middle and high latitudes. Since the equatorial latitudes are most nearly perpendicular to the incoming solar radiation, the equator is least affected by seasonal variations. According to British meteorologist Hubert Lamb, the equator receives approximately 2.5 times as much heat as the poles during the course of the year.

Variances within the constituents of the earth's atmosphere influence the balance of incoming and outgoing radiation. Changes in the concentration of such atmospheric components as aerosols or particles, ozone, carbon dioxide, or water vapor affect the earth's radiational balance. As the earth's heat balance adjusts to the altering equilibrium, it affects the mean surface temperature and climate conditions.

Another important factor in the influence of the radiational balance is the surface-atmosphere system of the earth, and the reflectivity of the individual elements of this system. The proportion of reflectivity of solar radiation from the earth's surface-atmosphere system is known as the albedo of the earth. Approximately 30 to 35% of incoming solar energy is reflected back toward space, and comprises part of the outging radiation. The planetary albedo of 30 to 35% for the earth can vary because of the changes in the reflectivity of these elements. Anyone who has been outdoors on a sunny day in a snow-covered environment can attest to the high reflectivity of snow.

Depending upon the conditions of the surface and the age of the snow cover, snow and ice can reflect 30 to 80% of incoming solar energy. A coniferous forest covered with snow might reflect about 35% of incoming solar energy, while polar regions might reflect as much as 90 to 98%. Heavy cloud cover also has a high reflectivity. While thin, wispy clouds may have an insignificant reflectivity, thick stratus clouds can reflect about 70% of the incoming solar energy, and thunderclouds about 90%. Clouds play a significant role in the earth's albedo. With use of climate models, it has been calculated that without clouds the planetary albedo of the earth would be only about 20%. Rock and sand surfaces reflect about 15 to 30% of the solar energy, and vegetated land has a reflectivity of 10 to 20%, although the polar seas have a reflectivity as high as 25%, ocean water generally has the lowest reflectivity, with 5 to 10%. Instead of reflecting solar energy, oceans absorb the energy and act as heat reservoirs.

If one minor aspect of any element of the earth surface-atmosphere system undergoes change, the entire radiational balance is affected. We are just beginning to understand the complexity of the earth's atmosphere and its weather system as

159

a dynamic, constantly varying process seeking a transient equilibrium. It's the variable nature of the diverse elements within this system that drives our weather machine. In the final analysis, the earth's weather system is a vibrant process which provides vitality to the environment, allowing life forms to thrive.

The imbalance in heat received at different latitudes generates the earth's weather system through the circulation of the atmosphere and the ocean currents. As the equator receives more energy than do the polar regions, heat is transported from the equator towards the poles. At the poles, this convection continues, but now cooler air and water move toward the equator. In the convection cycle of energy that travels between the equator and the poles, one-third of the energy is transported by the ocean currents. Another third is transported by the direct transport of sensible heat by air circulation. The final third is transported by the latent heat of water vapor through the interaction of the atmosphere.

Much of the solar energy reaching the earth's surface is absorbed by the oceans, which act as heat reservoirs and serve as a moderating influence on temperature extremes. The diurnal and seasonal temperature ranges of oceans are far less than those of land masses. During the day and in the summer season, land becomes warmer than the oceans at the same latitude. During the night and in the winter season, the land becomes cooler than the oceans. The oceans have a tempering effect on the southern hemisphere. The 19% of continental land mass in this area has a more moderate annual temperature range than does the northern hemisphere, whose area is 39% continental land mass. While the average annual temperature range in the southern hemisphere is 7° Celsius, the average annual temperature range is 14° Celsius in the northern hemisphere.

The circulation of the ocean currents is influenced by the rotation of the earth, wind patterns, and differences in temperature and salinity within the oceans. The temperature and salinity of the ocean water determines the density of the water, the density being inversely proportional to the temperature, and inceasing with salinity. At the equator, the water warmed by the sun expands, raising the sea level and causing a flow toward the polar regions. Evaporation at the equatorial latitudes causes relatively high salinity in the warm water. As it nears the poles, the water from the warmer latitudes initially displaces the colder polar water before cooling. When the cooling water reaches a temperature comparable to that of the polar water, it becomes denser than the polar water because of its higher salinity. The warmer waters sink, causing an upwelling of the polar water it has displaced. Cooler, and lower in salinity, the polar water then moves toward the equator to repeat the cycle.

As the warm equatorial water moves toward the polar regions, the earth's rotation deflects the currents clockwise in the northern hemisphere, and counterclockwise in the southern hemisphere. The ocean currents are pushed westward by the easterly trade winds, and move toward the poles along the western side of the oceans. Conversely, they move along the eastern side of the oceans when they travel away from the poles.

The direct transport of sensible heat by air circulation accounts for another third of the energy that travels between the poles during the convection cycle, and functions by means of the atmospheric heat pump. Vertical convection is set in motion as a result of the inequality in the heating of the earth's surface and the troposphere's characteristic of temperature decrease with height. Warm air at the equator rises and moves toward the poles. Cooler air at the higher latitudes descends and moves equatorward. This convection process was first postulated in 1735 by George Hadley, a London lawyer. Regularity in this circulation pattern occurs within the tropical wind regime between the equator and the 30° north and south latitudes. With one ''cell'' over each of the northern and southern hemispheres, these tropical cells, now known as the Hadley cells, tend to follow by one to two months the seasonal migration of the sun northward and southward. The Hadley cell pattern extends beyond the 30° latitudes during the summer-season in each of these two hemispheres.

When we consider surface wind circulation, we are not dealing solely with the concept of heat transfer from equatorial latitudes to higher latitudes and the return flow of winds back toward the tropics. Because our earth rotates, the wind patterns are deflected sideways. Together, these two phenomena produce spiral surface winds associated with low-pressure systems (cyclonic disturbances) and high-pressure systems (anticyclones). High-pressure systems form where cool air descends. The winds of a high-pressure system spiral outward. In the northern hemisphere, these winds rotate in a clockwise direction, but in the southern hemisphere, in a counterclockwise direction. Low-pressure systems form where warm air expands and rises. The winds of a low-pressure system rotate in a counterclockwise direction in the northern hemisphere, and clockwise in the southern hemisphere.

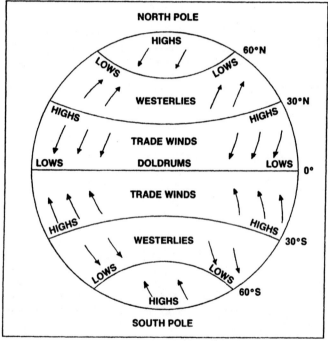

The patterns of surface winds and high- and low-pressure zones.

The prevailing pattern of surface winds is depicted in Diagram 12. The winds tend to blow from high-pressure regions to low-pressure zones. Between the equator and 30° latitude, the return flow of the Hadley cell is what we commonly refer to as the trade winds. Noted for their regularity, the trade winds are easterly, blowing from the northeast in the northern hemisphere and from the southeast in the southern hemisphere. When the trade winds come together near the equator, they become light and variable and are known as the doldrums, a low-pressure region where the air is rising. The doldrums are characterized by calms and heavy rains.

Around 30° latitude are the horse latitudes, a high-pressure region which borders the easterly winds of the trades and the westerly winds of the westerlies. Located between the 30° and 60° latitudes, the westerlies are variable winds because of migrating high-pressure and low-pressure systems. Although both the speed and direction of these winds tend to vary considerably, their average directional flow is from the west. At about 60° latitude is the low-pressure region of the polar front, which borders the westerly winds and the weak easterly winds of the polar latitudes.

While the surface wind patterns tend to alternate between regions of easterly and westerly flow, the jet streams of the upper air are westerly. The jet streams, tunnels of air that reach speeds of up to 400 kilometers an hour, travel eastward around the earth at altitudes of 10 to 50 kilometers. Jet streams are associated with the belts of high-pressure and low-pressure systems. Although there can be as many as seven jet streams during the course of the year, the main jet stream follows the path of the polar front at 60° latitude (allowing for latitude shifts corresponding to the seasonal migrations of the sun). The polar front jet stream, the first of its kind to be discovered, undulates in a zigzag pattern, bringing warm air northward and cool air southward. Changes in the velocity of the jet stream trigger the migrating high-pressure and low-pressure systems that contribute in the lower air to the irregularity between the latitudes of 30° and 60°.

Within the general patterns of the surface winds, regularity is disturbed by the surface features of the earth. If the earth were a smooth, uniform surface, instead of a collage of land masses and oceans, mountains, and plains, the wind patterns would be more regular. This irregularity accounts for a local influence on the prevailing winds of each region. An example of local influence can be seen in mountain ranges of the Rockies in North America. The north-south alignment of the Rockies

163

interrupts the westerly flow of the prevailing winds, forcing them to move north as they cross the mountains, then south as they move past the mountains.

The final one-third of energy that travels between the equator and the poles is transported by the latent heat of water vapor through the interaction of the atmosphere. The water vapor cycle operates through four basic steps: (1) evaporation, (2) transportation, (3) condensation, and (4) precipitation.

Because warm water evaporates more quickly than cold water, the greatest amount of precipitation occurs near the equator, where the water is warmer. The water evaporates into vapor and rises into the atmosphere. After its upward movement into the atmosphere, where the temperature decreases with height, the vapor cools. When it cools below its dew point, the vapor condenses into water drops, forms clouds, and causes precipitation. As already noted, much of the annual precipitation occurs in the equatorial latitudes. Some of the water that evaporates in the lower latitudes of the earth's surface rises and is transported toward the poles. Condensation at the poles can cause clouds, without precipitation. Latent heat, which originally served as the catalyst in changing water into vapor, is released during condensation. Through the interaction of the oceans and the atmosphere, latent heat is transferred from the lower equatorial latitudes toward the poles.

It's clear that there are numerous variables within the convection cycle that modify or distort the weather system. These changes produce both short-term and long-term climate fluctuations. In an attempt to determine the cycles of weather variations, climatologists have suggested cycles of approximately 2, 11, 22, 80, 180, 200, and 400 years.

It's important to recognize that statistical records for climate indices go back only 100 years for most recording stations, although a few have records which go back 200 years. In-depth statistical observations of atmospheric circulation are even more limited, taking into account conditions over the past 25 years. We might wonder, therefore, how climatologists can even describe cycles of climate fluctuations when the statistical evidence is so limited. The answer lies in the fact that weather conditions are recorded by our earth: in trees, in lake sediment, in ice layers.

The belief that trees hold a record of the earth's weather

was first proposed by Andrew Douglass in 1901. Douglass, a solar astronomer at the University of Arizona, felt he could discover indications of the sun's effects on weather as recorded by vegetation directly affected by that weather. Since the growth of a tree depends upon local temperature and precipitation, the annual growth recorded in the width of the rings in a tree indicates the local climate conditions for each year of the tree's life. Rings at the center are the oldest, and the rings nearest the tree's bark are the most recent. While local climate conditions can be determined by the width of the rings, their chemical content gives evidence of solar variations. Like all vegetation, trees absorb atmospheric carbon dioxide through photosynthesis. The amount of the isotope carbon-14 in the carbon dioxide contained in the tree rings provides a record of the sun's activity. During the sun's active periods less carbon-14 is formed in the upper atmosphere and consequently less will be found in the carbon dioxide in the tree rings. When the sun is quieter, more carbon-14 is formed in the upper atmosphere, and more of it will be found in the tree rings.

Sediment deposits in lakes also provide information about climate conditions. By analyzing the layers upon layers of sediment on lake bottoms, scientists can determine the state of the vegetation and the soil content in past years. Through the work of Reid Bryson of the University of Wisconsin and his colleagues, this study of lake varves (layers of lake sediment) has provided records of local climate changes over the last 10,000 years.

A third technique to discover past climate conditions is the study of ice core samples. As with lake varves, the ice layers near the top represent the recent past and those toward the bottom the distant past. One of the most successful expeditions for the drilling of ice core samples began in 1966 at Camp Century in northwestern Greenland. One core taken from the Greenland ice is believed to date back 150,000 years. Willi Dansgaard of the University of Copenhagen and others have analyzed samples for their relative content of the heavy oxygen atom, oxygen-18, and the lighter oxygen atom, oxygen-16. More oxygen-18 atom indicates that the climate was warm; less oxygen-18 atom denotes a cooler period. Based on his studies of the ice cores, Dansgaard found regular cycles of 80 and 180 years in climate conditions which he feels are related to solar variations. The question arises whether solar activity could have such a dramatic impact on earth weather conditions as to precipitate the ice ages in the past. To answer this, we must speculate on the causes of large-scale climate fluctuations, which can be attributed to three general and interrelated categories:

I. Events associated with the earth's relationship to the solar system.
II. Events associated with the processes occurring within the earth's system.
III. Events associated with human activities.

Focusing their attention upon the earth's position relative to the sun, scientists see three primary variations to the earth's orbital parameters. These three variations are termed: (1) the precession of the equinoxes, (2) the obliquity of the ecliptic, and (3) the eccentricity of the orbit.

The precession of the equinoxes is a cycle of approximately 21,000 to 26,000 years. Because of the gravitational influence of the moon and the sun, our ellipsoid-shaped earth wobbles as it revolves around the sun. If we were to extend an imaginary line from the earth's axis out into space, it would describe a circle in space similar to the movement of a dying spinning top. This wobbling of the earth determines whether its northern or southern hemisphere receives the greatest impact of solar radiation when it is closest to the sun. At present, the earth is closest to the sun in January, during the southern hemisphere's summer. Some 10,000 years ago the earth came closest to the sun in July. When the earth is furthest from the sun during the summer season, it promotes cooler temperatures. Consequently, summers in the northern hemisphere are now cooler than summers in the southern hemisphere.

The obliquity of the ecliptic is a cycle of approximately 40,000 to 41,000 years. The angle of incoming solar radiation varies according to the changing tilt of the earth relative to the sun. This phenomenon causes the seasonal variations we experience. The rolling tilt of the earth's rotation axis to its orbital plane can range from angles of 21.8° to 24.5°. The degree of tilt affects the seasonal extremes, for the greater the tilt the

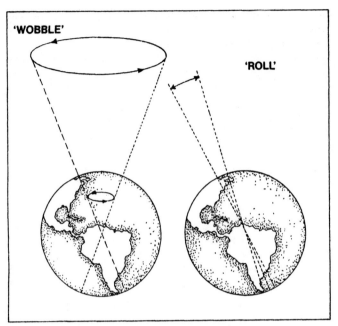

The earth's wobble causes the apparent precession
of the equinoxes.

The earth's roll, brought about by the obliquity of the ecliptic.

more pronounced the differences between summer and winter.
The last maximal tilt occurred about 10,000 years ago, and the
angle of tilt is now about 23.5°. Taken alone, this lessening of
the angle of tilt would lead to more moderate seasonal variations, to cooler summers and warmer winters.

The eccentricity of the earth's orbit is a cycle of approximately 90,000 to 100,000 years. The earth's orbit around the
sun varies from the nearly circular to slightly elliptical and
back to nearly circular. Over the last four million years the
eccentricity of the earth's orbit is calculated to have ranged
from 0 to 0.07. While this stretching of the earth's orbit might

seem insignificant, it, too, affects the amount of solar radiation that reaches us. When the orbit is stretched to its maximum, the intensity of solar radiation can vary up to 30% from its strongest to its weakest during the course of the year. At present, this annual variance in solar radiation is about 7%.

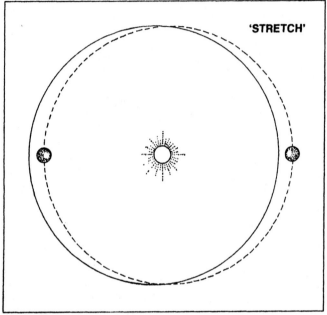

'STRETCH'

The earth's stretch, brought about by the eccentricity of the earth's orbit.

These three variations in the earth's position relative to the sun are known collectively as the Milankovitch effect, named after the Serbian geophysicist, Milutin Milankovitch. In the 1920's, and 1930's, Milankovitch formulated a detailed model of the earth's orbital variations and postulated their effects. One effect of the changes in the earth's orbital parameters that

Milankovitch suggested was the onslaught of the ice ages. In 1976, impressive evidence was found to support him. A team of scientists, led by James D. Hays of Columbia University, John Imbrie of Brown University, and Nicholas Shackleton of Cambridge University discovered supporting evidence in sediment cores taken from undisturbed sections of the Indian Ocean. These cores provided a complete record of climate changes over a period of 450,000 years. Within them were remains of the plankton radiolaria, which is especially sensitive to temperature change. With such aid, the Hays team was able to estimate water temperatures of the Indian Ocean at different time periods. They found cycles of 21,000 years, 42,000 years, and 100,000 years—cycles associated with the three orbital parameter variations of the Milankovitch effect.

When Wegener proposed his theory of continental drift, it was met with hostility by those who believed, or needed to believe, in the permanency of continents and oceans. This same attitude holds true with regard to solar emission. Man believes solar emission to be constant.

The amount of solar energy received in the vicinity of the earth is termed the solar constant, which is given as 1.95 calories per square centimeter per minute. Before we accept this concept of solar constancy, however, let's keep in mind that the period of our observation has been limited. Although we have carefully observed the sun's activity for 200 years, the data from such a relatively brief period is insufficient to permit an accurate judgement of solar emission over millions of years.

During this century, man has estimated possible fractional variations of 1 to 2% in the solar constant. From the late 1800's until his death in 1973 at the age of 101, Charles Greeley Abbot, director of the Smithsonian Institution's Astro-Physical Laboratory, actually measured the sun's heat and light. Abbot believed that the solar constant was not constant, at all, but varied in regular 11-year and 22-year cycles, the sunspot cycles. At times of sunspot minimum, Abbot claimed the solar constant to be 1 to 2% less than at periods of sunspot activity. The Russian scientists, K. Ya. Kondratyev, and G.A. Nikolsky assert that the solar constant increases with sunspot activity until it reaches a maximal variation, after which the variation decreases. According to their findings, the solar constant is approximately 2% higher during moderate sunspot activity than during no such activity. This variation decreases when sunspot activity is high.

Not all scientists are convinced that this relationship exists. Some believe that the sunspot cycle effect on the solar constant is negligible. While the sun's surface temperature and the solar constant may appear to change, these scientists emphasize that this apparent change is due to the phenomenon known as "line-blanketing," and not to a change in solar emission. Line-blanketing shifts the distribution of the sun's photon output from one end of the spectrum to the other. A change occurs in the wavelengths of the sun's emission, but not necessarily in the amount of emission. When there is sunspot activity, the spectrum of incoming radiation changes with an increase in ultraviolet.

Jack Eddy, of the University of Colorado's Department of Astro-Geophysics, is one scientist who believes that the solar constant not only varies in accord with the sunspot cycle but also varies as an inherent characteristic of the sun. From his analysis of Abbot's measurements, Eddy saw the regular variations of the sun's energy emission associated with the sunspot cycle. When he presented his findings in 1977, he brought out something that Abbot himself had missed. In the long-term averages of Abbot's measurements, the solar emission appears to be inceasing over time. Calculating a rise of about .5% per century, Eddy suggests that the sun is not constant, but rather a variable star which is getting hotter. Such a suggestion might seem radical, but it has been proposed by other scientists as well.

Although it may be years before the scientific establishment accepts the sun as being a variable star, evidence increasingly supports the periodic variations in its energy associated with the sunspot cycle. From earth, we see sunspots as dark blemishes on the sun. Usually seen in pairs, sunspots are cooler areas of the sun's surface where whirling loops of gas break through. Since these gas loops follow lines of magnetic force, sunspots have strong magnetic fields.

According to present solar theory, sunspot activity is caused by the differential rotation and the convective motion of the sun. The sun does not rotate like a solid body, but with a lag at its polar latitudes. While the sun's equator rotates in 27 days, its latitudes do so in 31 days. Within the sun are magnetic forces that interact with the differential rotation to generate electric currents. Through the process of convection, gas columns rise towards the sun's surface and fall back toward its center. During sunspot activity, these gases break through the surface at areas of strong magnetic force. Bursts of ionized gas spew forth into the universal atmosphere.

Most numerous during times of sunspot maximum, these solar flares send strong doses of ultraviolet, charged particles, and X-rays out into interstellar space. These flares follow the solar wind, a gaseous continuation of the solar atmosphere that streams electrically-charged particles outward in all directions. Near the earth, this stream of charged particles is funneled down into our polar regions by the geomagnetic field. The earth's magnetic field is affected by increased solar activity. There is a greater incidence of both the aurora borealis (northern lights) and the aurora australis (southern lights). Communication systems, such as television, short-wave radio, and citizen's bands are disrupted. Surges of electricity pass through power transmission lines, threatening transformers and causing electrical blackouts.

Spots on the sun were noted as early as the fifth century B.C., but not until the mid-nineteenth century A.D. was it realized that they appeared in regular cycles. A German amateur astronomer named Heinrich Schwabe recorded the sunspots he observed, and plotted their average number per year. By 1843, he believed he had discovered a regular cycle to the activity. After recording two of the cycles, Schwabe announced in 1851 the discovery of a consistent sunspot cycle. Although he felt that it recurred over a period of 10 years, this periodicity was later refined to 11.2 years.

Similar to crests and troughs in the ocean waves, the sunspot cycle proceeds from one period of maximal sunspot activity down to a minimum, and then up to the next maximum. Although the cycle averages 11.2 years, individual cycles can range from seven to 18 years. During one cycle, sunspots with a positive polarity may lead in the sun's northern hemisphere, while those in the southern hemisphere will have a negative polarity. During the following sunspot cycle, that order will be reversed. This phenomenon of alternating magnetic polarities associated with the 11-year sunspot cycles has been observed since 1908 and has led to the cycle known as the 22-year Hale magnetic cycle—two 11-year cycles which bring a return to the same starting polarity.

These rhythmic 11-year, 22-year sunspot cycles, however, can be distorted by planetary configurations and their consequent influence on the sun. John H. Nelson (cf. Chapter II) in his study for RCA found that solar storms were intensified by planetary alignment to the sun in angular relationships of 0°, 90°, and 180°. In Chapter VII we noted the belief of John Gribbin and Stephen Plagemann that the 179-year cycle, when all nine planets move into an alignment with the sun, not only intensifies the magnetic storms on the sun but may affect the earth's atmosphere, disrupt weather patterns, and even trigger catastrophic earthquakes. Within this 179-year cycle are subperiods of 80 and 100 years. These various cycles of planetary interactions in relationship to the sun are fascinating in light of the conclusions reached by Willi Dansgaard from his studies of ice cores that climate conditions go through regular cycles of 80 years and 180 years.

If we take into account the regular sunspot cycle, the distortion of solar activity by planetary configurations, and the likelihood of the sun's being a variable star, we have the ingredients for fluctuations in solar energy emission with potentially radical consequences for earth conditions. Skeptics who continue to scoff at the influences of a vibrational universe, constantly changing, continuously exerting subtle influences upon our earth. Beware! Evidence of solar variations and their influences on earth conditions is mounting. We know that sunspot cycles can and do affect our earth.

From the studies of climate records contained in tree rings, lake varves, and ice cores, scientists see cooling trends associated with major decreases in the sun's activity and warming trends with the major increases. An example of this correlation is seen in the period from 1645 to 1715, a time when the sun seems to have lost its spots. E. Walter Maunder, a nineteenth-century English astronomer, discovered that sunspots were rare during those 70 years. Not only had they virtually vanished, but the coronae usually seen around the sun during solar eclipses were no longer seen. Neither the aurora borealis nor the aurora australis were seen during this period. Now known as the Maunder Minimum, this period of radically reduced solar activity coincides with the coldest interval known in western history.

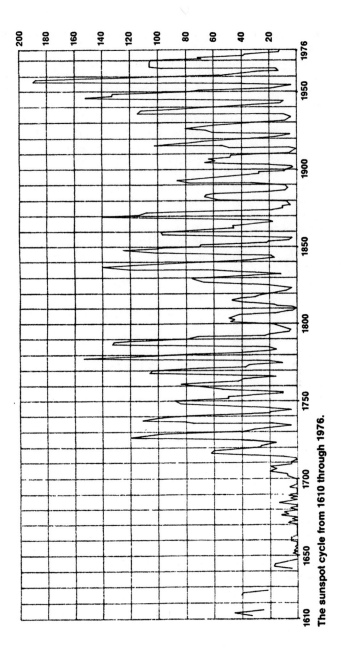

The sunspot cycle from 1610 through 1976.

It's referred to as the Little Ice Age, a time of lowered temperatures, glacier advance, and widespread hunger. This correlation has been confirmed by measurements of the carbon-14 in tree rings of the period, which during those 70 years was significantly higher than usual, indicating that solar activity was remarkably quiet.

A recurring correlation between the 22-year sunspot cycle and earth conditions is evident in the cycles of drought that strike the western plains of the United States. From analyses dating back to 1700 and taken from 40 to 65 different sites west of the Mississippi River, Charles Stockton and David Meko, both of the University of Arizona, discovered a cyclical shrinking and swelling of tree ring growth. This cyclical variation revealed that the plains states had gone through recurring periods of drought on an average of every 22 years. Carrying these studies further, Murray Mitchell of the National Oceanic and Atmospheric Administration, found the timing of the drought to be dependent on the strength of the sunspot cycle. Strength was determined by the total number of sunspots during the cycle. Mitchell asserts that when the drought occurs during a weak sunspot cycle, the maximal area of drought coincides with the sunspot minimum. If the drought takes place during a strong sunspot cycle, the maximal area of drought follows the sunspot minimum by two to three years.

This notorious drought was immortalized in John Steinbeck's novel, *The Grapes of Wrath*, and etched deeply into the memories of those who lived in the American plains states in the 1930's. Beginning in 1930, the worst drought in American history struck. By 1935, the south central states were referred to as the "Dust Bowl." The drought extended across one-fifth of the American land mass and was far more severe than the droughts that had hit 20 years earlier or 20 years later.

In 1976, a time of sunspot minimum, drought again threatened the western plains. Rain did finally come by late spring, but the harvest of 1976 was still down 10 to 15% from the previous year. That same year, California, the major recipient of immigrants from the Dust Bowl of the 1930's, suffered a major drought. During 1976 and 1977, California experienced historic lows in rainfall, forcing severe water rationing until the drought broke in early 1978. While the pattern in the mid-1970's occurred further west than usual, the drought still came.

Since warmer weather appears to be associated with increased solar activity, it stands to reason that food production and agricultural harvests are affected by sunspot activity. This is especially true in the northern hemisphere, where the growing season tends to be about 25 days longer during periods of sunspot maximum than during periods of sunspot minimum. As an example, the wheat harvest in 1968, a year of sunspot maximum, was greater than the annual harvests for the following four years. Sometimes the maximal harvest yield during a sunspot cycle follows the sunspot maximum by a year, which was the case in 1958. During that cycle, the year of sunspot maximum was 1957; the 1958 wheat harvest was greater than those for the following five years. Conversely, 1954 was a year of sunspot minimum, and yielded a disappointing harvest. Wheat production was less than the annual yield for the two preceding and the two succeeding years. Although this correlation between sunspot maxima and good harvests seems true for the northern hemisphere, figures for the southern hemisphere defy this generalization, and point instead to the reverse phenomenon. Yields in southern hemisphere countries are usually better at times of sunspot minimum. Nonetheless, global averages do appear to bear out the correlation between longer growing seasons and higher sunspot activity.

While sunspot activity seems to benefit global temperature averages, it can have adverse consequences by depleting the ozone in the stratosphere. Paul J. Crutzen, of the National Oceanic adt Atmospheric Administration, has studied the rhythmic 11-year variations in the atmospheric ozone concentrations and has proposed the process by which solar activity destroys ozone. Electrically-charged particles emitted by solar flares disassociate nitrogen (N_2) and cause nitric oxide (NO) to form. The nitric oxide then reacts with the ozone. Since one nitrogen oxide molecule can react with 100,000 ozone molecules, Crutzen believes that even a small injection of nitric oxide can have a significant impact on the ozone layer. Confirmations of Crutzen's theory came from the effects of the sun's activity during the first week of August 1972, when nine major flares erupted on the sun. As a result of these massive

175

eruptions, Crutzen predicted an ozone decrease of 15 to 20% over the North Pole. During the eruptions, a Nimbus 4 weather satellite was in orbit, measuring and recording the ozone concentrations. From their analysis of the recorded data, Donald Heath and Arlin Krueger, both of the Goddard Space Flight Center, found that the ozone had been depleted by 16%, a figure consistent with Crutzen's predicted range.

Because the ozone shields the earth from the ultraviolet rays of solar radiation, its depletion can lead to higher and potentially lethal concentrations of ultraviolet reaching the earth. Periods of sunspot maximum can thus pose the danger of high doses of radiation to astronauts and passengers of high-flying aircraft. Incidences of melanoma, a skin cancer, are also likely to rise at such times. During the most recent sunspot cycle, the sunspot minimum occurred in 1976 and is forecast to peak around 1982.

The sun's variability and the earth's relationship to the solar system are not the only sources of major climate changes. Plate tectonics and the internal variability of the earth's weather system are also important factors in climate changes.

Throughout this book, we have consistently looked at the earth as a living, breathing entity which is constantly changing. The dynamics involved in these changes affect the radiational balance between the absorption and reflection of solar radiation at the earth's surface. A case in point is volcanic eruption. As we saw in Chapter VI, volcanic eruptions and their consequent "dust veils" can produce sharp changes in global temperature, to the extent of creating the dire conditons of hunger and famine.

The "dust veils" filter the incoming solar radiation. Since the wavelengths of incoming solar radiation are either equal to or smaller than the volcanic dust particles, these particles will absorb, disperse, or reflect back some of the solar radiation. Although some loss of incoming radiation will be compensated for by the increase of diffuse radiation (scattered or dispersed light rays), the condition is aggravated by the inability of the volcanic dust particles to absorb or intercept the longer wavelengths of the outgoing radiation. Consequently, the level of outgoing radiation remains steady while the incoming radiation is reduced.

Based upon a mean value of direct solar radiation for the

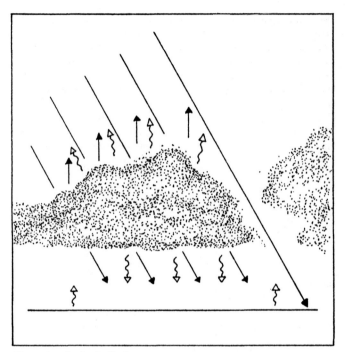

The volcanic dust effect.

years 1883 to 1938, scientists have found that after volcanic eruptions in the years 1883, 1888, 1902, and 1912, the monthly average of direct solar radiation dropped by as much as 20 to 22% below the mean. Volcanic dust veils can have an impact on incoming solar radiation for up to three years after eruption. Some of the particulate matter stays in the stratosphere for as long as 12 years after eruption. From his studies of the recorded volcanic eruptions between 1500 A.D. and the 1960's, Hubert Lamb concluded that major volcanic eruptions can indeed have a harsh, temporary effect on the world's climates.

Other scientists go even further, suggesting a correlation between volcanic activity and the earth's glacial ages. Injecting large quantities of dust and ash into the atmosphere, extended volcanic activity would prevent a certain amount of solar radiation from reaching the earth's surface and thus reduce the earth's surface temperatures. If maintained over a long time, the volcanic dust veil would severely disrupt the climatic equilibrium and might even lead to a spread of glaciation away from the poles. The correlation between volcanic activity and glaciation was proposed in 1974 by J.P. Kennett and R.C. Thunnell of the University of Rhode Island. From their analyses of ocean sediment cores, they concluded that the increased explosive volcanism during the Quaternary period triggered the succeeding Great Ice Age of the Pleistocene epoch (about one million years ago).

Volcanic particles injected into the atmosphere by eruptions are calculated to have increased by thirty times between 1961 and 1968. Should this trend continue, as expected, volcanic activity will influence the radiational balance and play a significant role in large-scale climate fluctuations.

Another of the dynamics of our earth is the internal variability of our weather system. Even the change in one element of the weather machine can have global consequences. Since the various elements of the earth's surface-atmosphere system have different qualities of reflection, minor transient variations in reflectivity conditions will also affect the radiational balance and the climate. For instance, although thin clouds have an insignificant reflectivity, heavy cloud cover can reflect 70% of incoming solar radiation. As the earth's albedo, or proportion of reflected solar radiation, would be about 20% without cloud cover, instead of the present 30 to 35%, it is obvious that changes in the concentration of cloud cover can have a severe impact on climate. Such is true of snow or ice cover. Snow and ice can reflect anywhere from 30 to 80% of the incoming solar radiation. Rock and sand surfaces have a reflectivity of about 15 to 30%, and vegetated land a reflectivity of about 10 to 20%. Assume that snow and ice cover these surfaces. The result would be an increase in reflectivity that would distort the radiational balance.

Pumping heat from the equatorial latitudes toward the polar regions and cold from the polar latitudes toward the equator, the weather machine depends upon air circulation, ocean currents, and the interaction between the atmosphere and the oceans for transport. Should any of these three be modified, not only will it have an impact on the other two, but more importantly, it will hinder the transport of heat and cold. In turn, this could intensify temperature extremes between different latitudes and affect climate conditions.

The following two chapters will lead us into a discussion of weather extremes. In the past, nature's own course has created dramatic climate fluctuations. Man's own influence as a contributor to large-scale climate fluctuations has become significant only during the present era.

X

FIRE AND ICE

Some say the world will end in fire,
Some say in ice.
From what I've tasted of desire
I hold with those who favor fire.
But if it had to perish twice, I think I know enough of
  hate
To say that for destruction ice
Is also great
And would suffice.

—Robert Frost

At different periods during the earth's history, its radiation balance shifted so radically that the severe cooling trends which resulted transformed the earth into an ice-laden planet. The eventual temperature drops, estimated at 6 to 8° Celsius, generated a spreading of ice sheets away from the polar regions over the continental land masses toward the lower latitudes, and inaugurated those periods known as the ice ages. Until the last 10 to 20 years, modern scientists accepted the belief that the earth had experienced four such ice ages during the Pleistocene epoch of the last million years.

Evidence of the four ice ages was first proposed in 1909 by Albrecht Penck of the University of Berlin and Eduard Brückner of the University of Vienna in their work, *Die Alpen Im Eiszeitalter.* According to Penck and Brückner each of the four ice ages of the Pleistocene epoch lasted about 100,000 years and was separated by warm interglacial periods, lasting between 125,000 and 275,000 years. Ocean levels dropped approximately 100 to 145 meters during the ice ages and rose 15 to 80 meters during the interglacial periods. The four periods of Alpine glaciation were named by Penck and Brückner after Bavarian streams. We list these period here along with their corresponding North American glaciations:

Alpine	North American	Approximate Dates of Occurrence
Günz	Nebraskan	about 800,000 to 900,000 years ago
Mindel	Kansan	about 600,000 to 700,000 years ago
Riss	Illinoian	about 200,000 to 300,000 years ago
Würm	Wisconsin	about 15,000 to 60,000 years ago

This sequence met with immediate acceptance by the scientific community until 1955, when Cesare Emiliani, of the University of Miami found evidence in sediment cores taken from the Caribbean which indicated that there were eight rather than four glacial periods during the Pleistocene Age. According to this recent theory, the warm periods between glaciations lasted only 10,000 years, considerably shorter than the intervals suggested by the earlier theory. As the evidence supporting this theory is generally accepted by scientists today, there is a genuine concern for the present generation which, 15,000 years away from the last ice age, appears to be living on borrowed time before the next glacial period. Equally alarming is the finding of Willi Dansgaard, who through his study of ice cores, uncovered evidence that 90,000 years ago the earth, still in the midst of fairly warm climate conditions, was suddenly, within a 100-year period, chilled and cast into a 1000-year period of climate severity that rivalled the conditions of an ice age.

181

Based upon past climate records, we can no longer assume that the present interglacial period will last another 100,000 years. On the contrary, we must accept the fact that severely cold climate conditions may again besiege the earth, leading to the spread of ice cover across the planet, perhaps initiating another ice age. As the danger confronts us as real, theories proliferate as to the reasons for the onset of ice ages. Although these theories vary in their particulars, they all emphasize the reinforcing elements at work once the cooling process is set in motion.

One of these is the "snowtrap" theory. Since snow falling on snow-covered ground takes longer to melt than snow falling on uncovered ground, the snowtrap theory claims that a small area of ice or snow cover can serve as a snowtrap. Based upon observations of glacial spreading, the snowtrap theory holds that the initial snowtrap forms on high ground, such as mountain ranges. As more snow falls on the area, the snowtrap thickens and extends outward. The process eventually becomes self-perpetuating as the snowtrap becomes larger and creates ever more conducive conditions for a build-up and spreading of the ice cover.

While the "snowtrap" theory posits the horizontal extension of ice cover, the "snowblitz" theory posits the vertical build-up of an ice sheet. This theory, proposed in 1970 by Hubert Lamb and his associate Alastair Woodroffe as a result of their studies of atmospheric circulation patterns during the ice ages, contends that an anomalous winter season can provoke the growth of ice sheets. Whether by increased volcanic eruptions whose dust veils temporarily lower the earth's surface temperatures or by changes in the atmospheric circulation patterns, or by some other anomalous condition, the theory proposes that after a severe winter with heavy snowfall, snow cover might not totally melt during the summer. Because of the increased reflectivity of solar radiation, the continued snow cover lowers surface temperatures and provides the impetus for increased snowfall during the next winter. Again, the process becomes self-propelling as the greater snowfall of the second winter also does not melt in the summer and begins to establish permanent snowfields. The thickening snowfields lower the surface temperatures further, providing conditions conducive for more snow during the third winter and continued growth of the ice or snow cover.

The third is the "ice-surge" theory, proposed in 1964 by A. T. Wilson of Victoria University in New Zealand. Although some of the ice cover on Antarctica creeps toward the coastal edges and breaks off from the continent as icebergs, the ice-surge theory contends that a massive build-up of ice cover on that polar continent could prevent the heat rising from the earth's interior. The prevention of heat release then warms the bottom layers of the ice sheet, melting them and thereby lubricating the ice sheet, allowing it to slide relatively easily along the ground. This process eventually leads to the ice sheets breaking off into the ocean as large sections instead of as smaller icebergs. The immediate consequence of such an ice surge is the triggering of *tsunamis* and the raising of the level of the world's oceans by as much as 20 to 70 meters, thereby flooding many coastal areas around the world. In addition, it is asserted that an ice surge distorts the earth's radiational balance, for the huge block of floating ice increases the amount of solar radiation reflected back toward interstellar space. This condition in turn results in a dramatic lowering of surface temperatures, a cooling trend, and the beginning of another ice age.

In 1934, George C. Simpson of the Meteorological Office in London proposed a theory which explained how slight increases in solar radiation and reduced ice cover during relatively warm periods could initiate an ice age. According to Simpson, these two factors would intensify atmospheric circulation, which would thereby enhance moisture transport toward the higher latitudes and produce greater snowfalls in the polar regions. The greater snowfall, because of its increased density, would melt more slowly and eventually lead to the establishment of permanent snowfields. The self-propelling process of growing snow and ice cover could initiate the resurgence of another ice age.

While Lamb and Woodroffe's "snowblitz" theory presently appears to be the most credible explanation of the growth of ice ages, all these various theories agree on one fundamental point—minor, seemingly insignificant, changes can institute self-propelling conditions that could result in the severe conditions of an ice age.

A general scenario of the oscillations between warm and cool periods is outlined below. Let us assume that the earth is in a relatively warm period. As global temperatures become increasingly moderate, the polar ice caps will begin to melt. This melting will raise the level of the world's oceans, and also allow the warmer water of the lower, equatorial regions to rise to higher latitudes. At the same time, the melting of the ice cover and the interaction between the warmer water of the temperate latitudes and the colder polar water will lower the temperature of the waters returning to the lower latitudes. As this process occurs in the ocean currents, the water vapor cycle will intensify. Increased evaporation, condensation, cloud cover, and precipitation will result in greater snowfalls at the higher latitudes. Cooler water temperatures at the lower latitudes and greater snowfalls at the higher latitudes will move the world into a cooling trend. This process will be augmented by the increasing reflectivity of solar radiation from both the greater cloud cover and the growing snow and ice cover. It is estimated that an increase of approximately 1% in low cloud cover over the entire earth could lower average earth surface temperatures 0.8 Celsius. And as we noted in the preceding chapter, snow and ice cover in the polar regions can reflect between 80 to 98% of the incoming solar energy, whereas oceans act more like heat reservoirs, absorbing the solar energy and reflecting only about 5 to 10% of the incoming solar radiation.

Although this cooling trend could continue to worsen, perhaps even to precipitate an ice age, it might also eventually reach a limit point which would provoke a switch to a warming trend. Such a point would be reached close to the point of stagnation in the momentum of ocean currents, atmospheric circulation, and the interaction of the oceans and atmosphere in the water vapor cycle. The now cooler waters of the lower equatorial-area latitudes would no longer transfer as much heat to the higher polar-latitude waters, many of which would be bottled up by the ice. Due to the cooler air temperatures at the lower latitudes, the global atmospheric circulation would weaken. And the aridity of the cooler phase would lead to a decrease in evaporation, condensation, cloud cover, and snowfall. As both the cloud cover and the snow and ice cover diminished, the high reflectivity of solar radiation would lessen, leading to increasing surface temperatures and the start of a warming trend.

On a smaller scale, this process can be seen in the seasonal fluctuations of the Arctic and Antarctic ice covers. The maximal amount of sea ice in the Arctic at the end of winter is about 5% of the northern hemisphere. The least amount of Arctic sea ice covers about 3.75% of the hemispheric area, producing an annual variation in Arctic sea ice of about 25%. In the Antarctic, the maximal amount of sea ice covers about 8% of the southern hemisphere. The least about 1.6%, resulting in an annual variation of about 75%. As the annual variation is far greater around the Antarctic than the Arctic, the Antarctic would seem more sensitive to slight variations and consequently more of a potential trigger for global climate shifts. Until recently, this assertion would have run counter to accepted scientific opinions which placed the Arctic regions in the forefront of climate changes which would only gradually reach the Antarctic. From his studies of sediment cores taken from the southern Indian Ocean near Antarctica, however, James D. Hays contends that past history indicates the Antarctic is more sensitive to climate variations and initiates changes that are experienced several thousand years later in the Arctic.

Yet, of what concern to us in the twentieth century are climatic changes which will develop in the next thousand years? We are more interested in the immediate future, which attaches to our present condition. Are we heading, in this near future, toward a cool or a warm period, or will our climate remain stable? Before we can understand even our most immediate prospects, we must recognize that we are living during an interglacial period. Viewed from the perspective of the earth's history, we are experiencing an abnormal era. The twentieth century has been a period of optimal climate conditions. According to Reid Bryson, current bountiful agricultural conditions have been unequalled since the eleventh century. But what of man's past? It is only by understanding the very different conditions of this past that we can gain the perspective we need to gauge accurately the possibilities of the future.

At about 1000 to 1200 A.D. the peak of the warm period that had begun around 500 B.C. was reached. At that time, human migration and agricultural cultivation spread northward, culminating in the ascendancy of the Vikings. Those two centuries attained a climatic optimum in the west which has since been unmatched. In the 1300's, the transition from a warm to a cold mode began in the west with stormy and erratic climate conditions. By 1430, the western world was in the grip of the "Little Ice Age."

Lasting until about 1850, the Little Ice Age brought widespread suffering and hunger to the northern countries. The Viking colonies died out. Cultivation moved southward as the ice encroached on the northern farmlands. The Alpine glaciers advanced, and the ice boundaries spread from the Arctic into the Baltic and North Atlantic. Hubert Lamb estimates that between the thirteenth and seventeenth centuries, the average temperature level fell by close to 1.5° Celsius. The worst periods of the Little Ice Age were 1430 to 1470 and 1550 to 1750. The 25 years between 1575 and 1600 are believed to have been the driest in the past seven centuries. Along with other climatologists, Lamb suggests that the Little Ice Age was provoked by the increased frequency of volcanic eruptions with their consequent dust veils that filtered out solar radiation. As noted in the preceding chapter, Jack Eddy uncovered a correspondence between the coldest interval of the Little Ice Age and the Maunder Minimum, when there was a dearth of sunspot activity. A decrease in the amount of solar energy reaching the earth's surface, through a combination of decreased emission and increased dust veils, could easily instigate a radical cooling trend similar to the one which precipitated the Little Ice Age.

Between the 1880's and 1940, the earth experienced a warming trend. During this period the mean temperature level for the entire earth rose by about 0.5° Celsius. In the northern hemisphere it rose by about 1.1° Celsius and in Iceland by as much as 2° Celsius. In the Arctic, this warming trend was even more pronounced and produced a decrease in the extent of the Arctic ice cover by about 10% and a reduction in the ice thickness by about 33%. As a result, the mean sea level is calculated to have risen 0.2 meters by 1940. With the retreat of both the northern ice and the glaciers, the cultivation of land moved northward. Growing seasons became longer, as in England, where Hubert Lamb estimates they were lengthened by two to three weeks. An increased rainfall penetrated the continental interiors, including the countries of the Sahel on the southern edge of the Sahara Desert. The monsoons, on which virtually half the world population is dependent for survival, came with fortuitous regularity. With this general improvement in climate conditions providing the best agricultural conditions since the eleventh century, food production increased dramatically. It seemed that the problem of world hunger might soon be solved.

In the 1940's, however, the earth entered a cooler phase, which has continued to the present. Pronounced in the Arctic and high northern latitudes, this cooling trend is most evident in the northern hemisphere, where the mean temperature from 1940 to 1970 declined by an estimated 0.33° Celsius. According to Hubert Lamb, the agricultural growing season decreased by about two weeks between 1950 and 1970. The cooling has also led to drier conditions with less rainfall and recurrent failures of the monsoon. During the late 1960's and into the early 1970's, a severe drought struck the southern fringe of the Sahara; it broke in 1974, but not before its consequences had claimed the lives of hundreds of thousands in the Sahelian countries. Although some scientists believe that the cooling trend leveled out in the early 1970's and has reversed itself into a warming trend, evidence of ice cover seems to confirm the continuing, albeit erratic, conditions of a cooling phase.

Since 1940, the ice cover in the northern hemisphere has been on the increase. In 1967, using observation satellites, the National Oceanic and Atmospheric Administration in the United States began systematically to monitor and map the extent of snow and ice cover in the northern hemisphere. In 1971, its snow and ice cover increased by about 11%, from an area of 33 to 34 million square kilometers before 1971 to 37 to 38 million square kilometers in 1971. Starting in 1971, and continuing through 1973, snow and ice cover in the northern hemisphere fluctuated around this higher level. Not only was the area of snow and ice more extensive, but, year by year, it formed earlier and took longer to recede. Although the comparatively milder winters of the mid-1970's seemed to support those who believe the recent cooling trend has begun to reverse itself, the fact remains that snow cover over North America in February 1978 was the largest recorded for that month since systematic monitoring began in 1967. And in December 1978 snow covered a larger area of North America than in any previous December since 1967. If we have returned to a warming trend, the most recent winters are certainly anomalies. As an example, the winter of 1978-1979 produced the harshest December in

187

100 years in European Russia, the coldest winter in western Europe since at least the Second World War, and an extremely bitter winter in the central regions of North America.

Climate trends, both present and future, are a source of lively debate in the scientific community. The three different positions which are assumed represent the entire spectrum of possible views on the subject: we are moving toward a cooling trend, a warming trend, or a period of stability.

Emphasizing the cycles of past climate changes, those who predict a cooling trend note that our current interglacial period has lasted longer than any in the past. Whether because of increased global cloud cover, evident since the 1960's; an increase in particulate matter injected into the atmosphere from geophysical and human activities; and/or a decline in solar energy emission, they propose that the decrease in solar radiation reaching the earth's surface will cause a drop in surface temperatures and incite the self-propelling process toward a general cooling of climate conditions. As further evidence in support of this theory, the proponents point to the cooling of the last 40 years, and to the general weakening of both atmospheric and oceanic circulation in recent years.

The proponents of the warming trend theory point to the intervention of men and women within the natural climatic process as the fundamental cause for the temperature reversal which they predict. Although they admit that we are in the midst of a natural cooling trend, they feel that, by the turn of the twenty-first century, or perhaps as early as the 1980's, human activity will interrupt this natural process. In the next chapter we shall examine the influence of men and women on the climate and what it portends. For the moment, it is sufficient to note that it is human actions which bear the responsibility of upsetting the natural progression of the present cooling trend.

Proponents of the third view, that no pronounced changes are likely to occur, attribute this stability to the cancelling out of the two contrary influences—human and natural. But there are also those, included in this category, who consider the present climatic information insufficient to forecast the future, and speculate that future developments may cancel out past knowledge.

Before we can determine the likelihood of a cooling or a

warming trend, we must measure the possibile extent and the exact consequences of the influence of human activity on climatic conditions. For whatever the eventuality, it is nevertheless clear that a change is inevitable. Constancy is the only impossibility where such conflicting forces are at work. Stability is the hope of those who cling for security to the present they know, and not a theoretical projection of those who have examined the conditions of change. The question is not whether there will be a change, but what the consequences of change will be.

During a cooling trend, the ice would spread out from the polar region rendering navigation in the higher latitudes of the northern hemisphere treacherous, if not impossible. The world's ocean levels would fall, which might affect the patterns of fish migration and perhaps even threaten their survival. It is likely that there would be lower annual catches by the fishing industry, and thus a reduction of a major food source. Over the longer term, the dropping ocean levels and cooler temperatures would mean less evaporation, less precipitation, and more droughts. Deserts would expand at the expense of arable land and forests.

A cooling trend could cause snowfalls to occur earlier and take longer to melt. The areas of cultivable land would shrink. The growing season would contract, and the cooler days would retard crop growth. Instead of the present situation in Europe, where one hectare feeds three persons, one estimate reveals that a 3° Celsius temperature drop in Europe would produce a drop in productivity such that one hectare would yield food for only two people. According to Reid Bryson's calculations, a decrease of only 1° Celsius in the mean annual temperature at Akureyri, Iceland, would not only shorten the growing season there by about two weeks, but would also reduce the number of growing degree-days by about 27%. Bryson estimates that a 2.4° Celsius drop would shorten the growing season by about 40 days and reduce the number of growing degree-days by as much as 54%.

Encroaching ice boundaries, falling ocean levels, extended length of snow cover, and increased frequency of unseasonal frosts are not the only unfavorable side effects of a cooling trend. A cooling trend would also weaken atmospheric circulation and increase both the storminess and variability of the weather. Weakening atmospheric circulation would create more pressure centers, although they would probably be smaller and weaker. These weak centers would tend to stagnate over areas for long periods, producing extended periods of unseasonably cold or warm weather on one side of the pressure center and possibly the reverse conditions on the other side. While some areas might suffer severe drought, others might be subjected to extensive flooding. These conditions could also change from season to season. Consequently, particular areas might experience alternating extremes in both temperature and precipitation.

In general, however, drought presents the greatest threat to the populated world during a cooling phase. As the precipitation belts in Africa would move further south, drought would renew its attack on the people of the Sahelian countries and promote the spread of the Sahara southward. The monsoons would fail more frequently, causing droughts in India and southeast Asia, threatening virtually half the world's population with hunger or starvation. The needs of these people would have to be met through international relief from the grain-producing countries. But such a solution assumes that the grain-producing countries would be both willing and able to meet their needs. Based largely on the work of Reid Bryson, a 1974 CIA report entitled "Potential Implications of Trends in World Population, Food Production, and Climate" states that a return to climate conditions like those of the Little Ice Age would lead to a loss of much of the arable land used for grain production in the U.S.S.R., a reduction of 75% in Canada's grain export capacity, the inability of India to prevent one-quarter of its population from starvation, and the need of China to import as much as 50 million tons of grain a year to avoid famine. A cooling trend would confront the world population with decreased crop harvests, a depletion of grain reserves, and eventually widespread starvation. One estimate goes so far as to project, within 25 years of a pronounced cooling trend, the death from starvation of 500 million people.

Carried to its extreme, a significant cooling trend could lead to another ice age with extensive glaciation and severe drought. In his book, *The Weather Machine,* Nigel Calder lists the prospects for various countries during such a period; those countries in danger of extensive or complete glaciation are shown in Table 1. Based on the work of Rhodes W. Fairbridge of Columbia University, Calder's assessment of those countries in danger of severe drought are shown in Table 2.

TABLE 1
COUNTRIES IN DANGER OF EXTENSIVE
OR COMPLETE GLACIATION

Afghanistan	East Germany	Norway
Argentina	Finland	Poland
Australia	Greenland	Sikkim
Austria	Iceland	Sweden
Bhutan	Irish Republic	Switzerland
Canada	Liechtenstein	U.S.S.R.
Chile	Mexico	United Kingdom
China	Nepal	U.S.A.
Columbia	Netherlands	West Germany
Denmark	New Zealand	

TABLE 2
COUNTRIES IN DANGER OF SEVERE
DROUGHT CONDITIONS

Afghanistan	French Guiana	Nigeria
Angola	Ghana	Pakistan
Argentina	Guatemala	Paraguay
Australia	Guinea	Senegal
Bangladesh	Guyana	South Africa
Botswana	India	Surinam
Brazil	Indo-China	Tanzania
Cameroun	Kenya	Togo
Central African Republic	Mali	Upper Volta
Chad	Mauritania	Uruguay
China	Mexico	Zaire
Congo	Namibia	Zambia
Dahomey	Niger	Zimbabwe

191

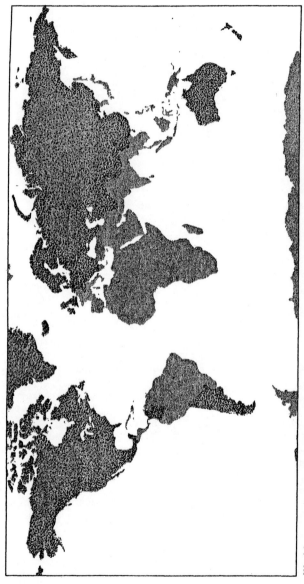

World map showing areas that would be affected by glaciation during a cooling trend.

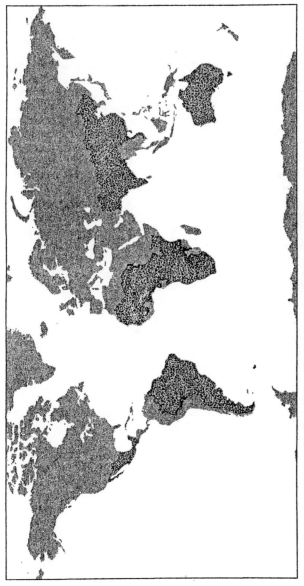

World map showing areas that would be affected by drought during a cooling trend.

If, on the other hand, the earth's climate is warming, the consequences of this change will be no less dramatic or problematic than those precipitated by a cooling. During a warming trend, the ice boundaries recede toward the higher latitudes, where the growing season would lengthen. Agronomists estimate that a temperature rise of 1° Celsius can cause a 10-day increase in the length of the growing season. Yet a 1° Celsius temperature change would have varying effects at different latitudes. Since a warming trend would intensify atmospheric circulation, the heat transfer from the equatorial latitudes poleward would accelerate. Consequently, an increase in the global mean temperature level by 1° Celsius might lead to minor increases of 0 to 0.5° Celsius at the equatorial latitudes between 10° N and 10° S, significant increases of 2 to 4° Celsius at the higher latitudes of 60° N and 60° S, and even more dramatic increases at the polar regions. In the middle higher and subtropical latitudes, the level of precipitation would be likely to increase. The monsoon-dependent lands would benefit from increased rainfall and there would be less likelihood of their failure. But the south-central regions of Eurasia and the central plains states of the United States would probably experience drought.

During both a warming and a cooling trend, some regions of the world would benefit, while others would suffer. Generally, however, the adherents of the warming trend theory contend that worldwide agricultural productivity would increase by as much as 50%, at least over the short term.

The warming trend is seen as largely fuelled by human activities, especially as they result in the increase of carbon dioxide in the atmosphere (a subject we discuss in the next chapter). Based on the climate model of Syukuro Manabe and Richard Wetherald, both of the Geophysical Fluid Dynamics Laboratory at Princeton University, the expected doubling of atmospheric carbon dioxide early in the twenty-first century is generally held responsible for producing an increase of 2.4° Celsius in the global mean temperature level. A substantial warming trend would trigger the melting of both the continental glaciers and polar ice sheets. Although the melting of sea ice would not raise the world's ocean levels, the melting of continental ice cover would. In fact, the greatest threat associated with a warming trend is a dramatic rise in the ocean levels and subsequent coastal flooding.

Carried to its extreme, a significant warming trend would melt the polar ice sheets, raising the world's ocean levels some 65 to 70 meters. *Hothouse Earth* by Howard A. Wilcox lists the prospects for various cities and countries during such a period. According to Wilcox, the larger cities likely to be flooded, probably flooded, or partially flooded are listed in Table 3. Table 4 indicates some cities that would not be flooded. Table 5 shows percentages of total land loss for the fifty states of the United States. Table 6 presents Wilcox's estimate of countries liable to lose 30% or more of their total land area to flooding.

TABLE 3
LARGE CITIES LIABLE TO FLOODING CONDITIONS

A. TOTALLY FLOODED

Bangkok	Karachi	Rio de Janeiro
Barcelona	Leningrad	Rome
Berlin	London	San Francisco
Bombay	Madras	Seoul
Boston	Manila	Shanghai
Buenos Aires	Melbourne	Sydney
Cairo	Montreal	Tientsin
Calcutta	Nagoya	Tokyo
Canton	New York	Victoria, Hong Kong
Djakarta	Osaka	Washington
Hamburg	Philadelphia	Wuhan

B. PROBABLY FLOODED
Peking

C. PARTIALLY FLOODED
Essen   Los Angeles   Manchester

TABLE 4
LARGE CITIES SAFE FROM FLOODING
CONDITIONS

Birmingham	Johannesburg	St. Louis
Budapest	Lima	Santiago
Chungking	Madrid	São Paulo
Chicago	Mexico City	Shenyang
Cleveland	Milan	Teheran
Delhi	Moscow	Vienna
Detroit	Paris	

TABLE 5
PERCENTAGES OF LAND LOSS FROM FLOODING
CONDITIONS IN THE FIFTY STATES OF
THE UNITED STATES

State	%of Land Loss	State	%of Land Loss
Alabama	40	Montana	0
Alaska	20	Nebraska	0
Arizona	0.5	Nevada	0
Arkansas	30	New Hampshire	20
California	10	New Jersey	40
Colorado	0	New Mexico	0
Connecticut	40	New York	10
Delaware	50 +	North Carolina	20
Florida	100	North Dakota	0
Georgia	40	Ohio	0
Hawaii	10	Oklahoma	0
Idaho	0	Oregon	10
Illinois	0	Pennsylvania	10
Indiana	0	Rhode Island	100
Iowa	0	South Carolina	40
Kansas	0	South Dakota	0
Kentucky	0	Tennessee	10

196

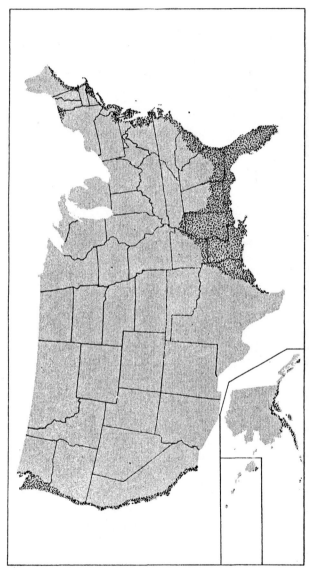

Map of the United States showing land areas susceptible to flooding and land loss during a warming trend.

State	%of Land Loss	State	%of Land Loss
Louisiana	100	Texas	20
Maine	20	Utah	0
Maryland	30	Vermont	10
Massachusetts	20	Virginia	30
Michigan	0	Washington	20
Minnesota	0	West Virginia	0
Mississippi	100	Wisconsin	0
Missouri	0	Wyoming	0

TABLE 6
COUNTRIES LIABLE TO LOSE 30% OR MORE OF THEIR TOTAL LAND AREA DUE TO FLOODING CONDITIONS

Country	%of Land Loss	Country	%of Land Loss
Australia	30	Malay Peninsula	40
Bangladesh	50+	Mozambique	30
Burma	30	Netherlands	100
Cambodia	40	Nicaragua	30
Cuba	40	Northern Ireland	30
Denmark	50+	Pakistan	30
East Germany	40	Paraguay	30
El Salvador	30	Philippines	30
England	30	Poland	40
Finland	40	Portuguese Guinea	50+
French Guiana	40	Senegal	50+
Ghana	30	Sierra Leone	30
Guyana	30	Sumatra	40
Hungary	30	Thailand	30
Indonesia	40	U.S.S.R.:	
Iraq	30	Western (20–90°E)	40
Irish Republic	30	Eastern (90–180°E)	30
Kuwait	40		

As we see in the projected scenarios for the two extremes of climate changes, either can bring adversity to the world population and trigger major changes to the surface features of the earth. The comparatively idyllic climate of the last hundred years has already begun to shift. The natural alterations of the cooling phase in progress have already begun to come into conflict with the climatic alterations produced by our industrial technology. It is on the terrain of this climatic battleground that men and women encounter their own responsibility for the outcome which is their own future.

XI

MAN'S FINGER AT THE TRIGGER

R: I've been thinking about what you said. About how small the span of human life is compared to the immensity of time. Of course you're right, but I am haunted by something Blake said—that although the generations of men are passed over by the tide of time, we still leave our traces which are permanent, for ever and ever. Doesn't this contradict your point?

L: Not at all. It is human knowledge which is limited; human power is as limitless as the universe. If we act wisely, we are capable of harmony as blissful as that of the mythical city of Andria.

R: I have never heard of this city. Please tell me about it.

L: It is a city which corresponds so perfectly with the skies that surround it, that one can scarcely make a distinction between them. Within the city, life glides regularly and calmly as the motion of celestial bodies. All of the inhabitants of Andria are self-confident and prudent. Because they believe that everything they do in their city has some effect on the sky which is their mirror reflection, before undertaking any activity they calculate the potential risks and benefits for themselves, their entire city, and all the worlds that exist beyond them.

R: But if we did not consider our actions as the people of Andria do, if, instead, we act recklessly, would our power still be limitless?

L: Certainly, but what we would create would be limitless destruction. We would be trading the utopia of Andria for the dystopia of Atlantis.

We arrive now at the third major cause of climate change: our own actions. The earth was around for a long time before *we* arrived. We've moved in and made ourselves at home, but it's time to realize that in many respects we've been ungrateful guests, or belligerent partners in one life-sustaining whole. Man's tampering with the natural balance of the earth can endanger not only himself and his loved ones, but can wipe out all living species. It's time to think about how to live *with* our changing earth. It's time to face reality—to recognize that we are gradually wearing thin our welcome. The cry of doom has sounded. Now it is time for constructive thought.

While the industrial revolution and accelerating technological advances have "improved" our life style, they have also played a major role in changing the earth's climate. Each day our actions affect variables that determine the earth's climate conditions. Virtually every move we make, from the turning on of a light switch to the irrigation of fields, affects our environment, and subtly influences the climate conditions we will experience tomorrow.

The theory which we examined in the last chapter, however, that these actions are simply countering the natural cooling trend in which we find ourselves at present, is not totally accurate. It is much too schematic, and does not pay adequate attention to the myriad factors which are involved in the process so summarily described as "countering."

In order to correct this error of simplicity, let us begin by examining those actions which fall under the category of land use. By altering the surface features of the earth, man affects the albedo (the proportion of solar radiation reflected back toward space), the surface resistance to air circulation, and the water-vapor cycle: evaporation, transportation, condensation, and precipitation. According to the Study of Man's Impact on Climate, a 1970 symposium sponsored by M.I.T., and hosted in Stockholm by the Royal Swedish Academy of Sciences, and the Royal Swedish Academy of Engineering Sciences, man's actions during the past 8000 years have dramatically transformed the land use of 18 to 20% of the continental areas. In recent years there has been an acceleration of this transformation.

The extension of cultivation and improper agricultural practices have a strong impact on local weather patterns with ramifications for global climate conditions. "Slash and burn" agricultural practices employed in the developing countries inject smoke and particulate matter into the atmosphere. As with a volcanic dust veil, the injection of particulate matter, or aerosols, into the atmosphere creates an aerosol veil that decreases the transparency of the atmosphere. This inevitably blocks some of the incoming solar radiation from reaching the earth, which then leads to a global cooling. Deforestation is practiced for many reasons, among them the expansion of grazing and cultivation, the use of firewood fuel in the developing countries, timber's value as an economic resource, and the spread of urban population centers. But many negative side effects also result.

Deforestation, for example, also injects particulate matter into the atmosphere. As forests are felled, the moisture content of the soil declines. The soil is then exposed to greater temperature extremes and is liable to severe erosion, either from heavy rains or from dry spells when winds sweep clouds of dust into the atmosphere. But the elimination of forests has other serious consequences for the earth's climate. Deforestation alters the radiational balance of a region, for it increases the reflectivity of the local surface area. Although deforestation in a tropical area might at first promote a warming trend, because of the high relative humidity, the eventual result of deforestation is temperature decline. The greater reflectivity of a deforested area means that more of the incoming solar radiation will be reflected back toward space than absorbed. A decrease in aerodynamic roughness, which results from the removal of trees, will intensify the wind-speed patterns. While stronger winds might accelerate evaporation, cooler temperatures, coupled with the lowered moisture content of the soil, will lead to less evaporation in the water-vapoi cycle and, consequently, less precipitation and less latent heat release. The region will then experience dryer, cooler weather than before deforestation.

Approximately 20 to 30% of the earth's total land area is covered by dense forest. But this percentage is sharply decreasing. Globally, forests are being cleared at an annual rate of 1 to 2%, roughly 11,000,000 hectares of forest lands a year. During the 1000 years from 900 A.D. to 1900, the forests of western Europe declined from more than 90% of the land area to just 18 to 23%. Although recent years have brought slight increases in the forest lands of Europe and North America, the developing countries in Africa, Asia, and Latin America are engaged in extensive deforestation. According to estimates based on the present rate, forest areas in developing countries will be reduced by about one-fourth by the turn of the century. Already, the tropical rain forests are only half as large as what they once were. By the year 2000 they may exist only as meager remnants. And the climatological consequences of this deforestation, as outlined above, could prove devastating.

While the United States is perceived as dedicated to the preservation of its forest lands, deforestation still takes place here as well. The United States Forest Service estimates that between 1962 and 1970 forest lands were cleared for cultivation at an annual rate of 200,000 hectares, or 2000 square kilometers a year. This does *not* include the forest land lost to encroaching urbanization, the construction of reservoirs, paving of highways, or the damage done by recreational use of forests. If these activities were included, the figure in the U.S. would be much higher. Yet the rate of deforestation in the United States is minor compared to that in other countries. For instance, J.P. Veillon studied the deforestation in Western Llanos, Venezuela, and discovered that during the twenty-five years between 1950 and 1975 there was a 32.5% decline in its forest.

Perhaps the most significant isolated case of deforestation is to be found in Brazil. In the state of Paraná, the 1953 forest area was 65,000 square kilometers, but over the next 10 years deforestation proceeded at a rate of 3% per year. With a total land area of 250,000 square kilometers, the state of São Paulo was 60% forest land in 1910. By 1950, this figure was reduced to 20%. But are we to fault the Brazilians for the deforestation of their land?

Brazil is often referred to as a "sleeping giant," for its wealth of natural resources make it potentially a giant in the world economic arena. Yet Brazil has been beset by various problems in trying to realize its potential. The major problem is the vastness of the country, the lack of a transportation network through its interior, and, consequently, the inaccessibility of the natural resources from the population centers along the coast. To develop that interior, the Brazilian government designed and built a new capital, the modern city of Brasilia, and also constructed the trans-Amazon highway, which is 2575 kilometers long. The highway, opened in the mid-1970's, has since been virtually abandoned as useless. Ecologists and meteorologists had protested loudly against its being built from the time it was first contemplated. They had good reason.

The Amazon Basin is the largest remaining forested area on earth. To slice even a small swath through it is to disrupt not only the ecological balance of the region but also the local climate, and to initiate potential consequences for global climate conditions. It has been alleged that the highway and consequent regional development of the Amazon could eliminate the entire jungle by the year 2050. Some estimates claim that as much as 20% of the area's timber has already been felled. In 1978, the Brazilian government announced plans to cut an additional 40 million hectares of trees in the Basin.

Recent reports from China offer evidence of the consequences to climate conditions caused by deforestation. In one mountainous region, deforestation is held responsible for a drop by one-half in the annual rainfall since 1967. During this period, wind-speed profiles in the area have increased from an average of 10 kilometers per hour to more than 60 kilometers per hour. In the tropical region of Yunnan Province, the loss of 133,000 hectares of forest land has led to hotter weather. One would expect the *initial* reaction to deforestation in a tropical region to be a warming trend because of the increase in the evaporation from the moist soil and in the release of latent heat from the water-vapor cycle. However, after the soil has dried out, the reduction in evaporation, precipitation, and latent heat release—combined with increased injection of dust into the atmosphere and greater reflectivity of the surface area—will most likely provoke a trend towards dryer, cooler weather.

Another impact of deforestation on climate results from the release of carbon dioxide into the atmosphere. Through the process of photosynthesis, vegetation absorbs carbon dioxide (CO_2) and water (H_2O) and converts them into carbohydrates (CH_2O) and oxygen (O_2). Laboratory studies have shown that increased levels of carbon dioxide stimulate photosynthesis. These studies, which monitored the carbon dioxide content of the atmosphere over a period of time, revealed a photosynthetic pulse that reflected a rise in the carbon dioxide content in the atmosphere during late winter when vegetation was dormant and a decline to a minimum during late summer when photosynthesis was at its peak. This evidence reveals that vegetation serves as a sink, or absorber, of carbon dioxide.

However, recent findings suggest that the flux in plant life may have the net effect of making it a source of carbon dioxide rather than a sink. The carbon stored in vegetation is released through deforestation, thereby contributing carbon dioxide to the atmosphere. This increase in carbon dioxide, along with that which results from the combustion of fossil fuels, can create a "greenhouse effect," and trigger a warming trend. The "greenhouse effect" is the effect of carbon dioxide on outgoing terrestrial radiation. Although carbon dioxide does not affect the incoming, short-wave solar radiation, it does block some of the outgoing, long-wave terrestrial radiation, some of which is absorbed and re-radiated back to earth. This process promotes a warming of the surface temperatures.

George Woodwell, director at the Ecosystems Center, Marine Biological Laboratory, at Woods Hole, Massachusetts, estimates the carbon contribution from deforestation is equal to 80 to 160% of the amount contributed by the combustion of fossil fuels. Other scientists see the input from these two sources as roughly equivalent.

Thus it appears that although the release of carbon dioxide into the atmosphere through deforestation would promote an initial warming trend, the long-term consequences are likely to be a trend toward dryer, cooler weather. Eventually, the deforested area would release no further carbon. Much of that already released would have found a sink and been removed from the atmosphere. The moisture content of the soil would have declined because of accelerated evaporation due to increased wind speeds. The evaporation rate would then drop, and the latent heat release from the water vapor cycle would fall. Tending to longer atmospheric life than carbon dioxide, the particulate matter injected by deforestation would contribute to the cooling trend, for aerosols generally absorb and radiate back to space the incoming solar radiation. The greater reflectivity of the earth's surface from deforestation would further alter the radiational balance, abetting the cooling trend.

Both in order to stem the damage caused by deforestation and to provide a cyclical crop of timber, governments and private corporations have been encouraging programs of reforestation. In Brazil, for example, reforestation in recent years has occurred on about 20% of the land that had been deforested in the state of São Paulo and on about 10% of the deforested land in Paraná. In some sections of the world, the abandonment of agriculture has actually led to the expansion of forests. In agricultural areas, belts of hedges and trees have reduced wind speed, decreased evaporation, and resulted in more bountiful crop yields. While efforts to reforest help repair some of the damage, the storage of carbon in reforested areas cannot match the carbon released from the felling of the primary forests they replace. Nor can it negate the damage already done.

Deforestation can foster the spread of deserts as well. In Brazil's northeast, the massive felling of trees has led to the intrusion of desertlike conditions on the once-humid region. Deserts can also be caused by other improper uses of the land, such as overplowing, overgrazing of livestock, and overharvesting of firewood. The plowing of fragile grasslands in the American plains contributed to the Dust Bowl during the long drought of the 1930's. The cutting of timber and overgrazing by settlers have degraded areas of Mexico and the southwestern United States, which are now deserts. Continued overgrazing in Australia may reduce three-fourths of that continent to desert. The overharvesting of firewood in the Sahel on the southern fringe of the Sahara is accelerating its spread—some 100 kilometers southward in the past twenty years.

At present, deserts cover 43% of the earth's land surface.

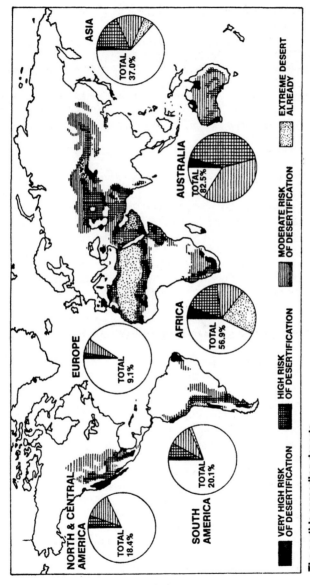

ASIA
TOTAL 37.0%

AUSTRALIA
TOTAL 82.5%

AFRICA
TOTAL 56.9%

EUROPE
TOTAL 9.1%

NORTH & CENTRAL AMERICA
TOTAL 18.4%

SOUTH AMERICA
TOTAL 20.1%

VERY HIGH RISK OF DESERTIFICATION

HIGH RISK OF DESERTIFICATION

MODERATE RISK OF DESERTIFICATION

EXTREME DESERT ALREADY

The earth's spreading deserts.

Approximately 14% of the world's population, roughly 630 million people, live on dry lands that offer only marginal subsistence. And the deserts continue to spread. Unless this spreading can be slowed, estimates suggest that up to one-third of the earth's present arable land will be lost to desert by the turn of the century. If we despair today over the suffering of our fellow human beings from malnutrition, hunger, and famine, we can only imagine the tragedy which will result when increased populations will make larger demands on ever smaller sources of food.

The spread of deserts also plays a significant role in contributing to a cooling trend. Like snow and ice sheets, deserts have a high reflectivity. The expansion of desert areas would alter the radiational balance by reflecting more of the incoming solar radiation back toward space. The injection of particulate matter into the atmosphere from the wind-swept desert sands would increase the atmospheric dust load; that aerosol veil would further diminish the amount of solar radiation reaching the earth. The higher reflectivity would promote the stagnation of air masses, effectively blocking the movement that might produce rain. The region's annual precipitation would decline, aided in part by the suppression of potential rainfall from the increased atmospheric dust load. Consequently, the advance of deserts would encourage a global trend toward dryer, cooler climate conditions.

While the purpose of irrigation is to bring water to dry soil and thereby increase its moisture content and productivity, irrigation has in some cases caused adverse effects and destroyed the land instead. Such a result can occur when irrigation waterlogs the land, leading to runoffs of the topsoil and severe erosion. Or it can occur when irrigation deposits toxic salts, as has happened in southern California. The construction of the Aswan Dam in Egypt allowed intensive irrigation of the Nile River Valley, but the consequences brought both waterlogging and salinization. A potentially catastrophic irrigation project, for other reasons, is the one contemplated by the Soviet Union to irrigate the farmlands of Siberia. The plan suggests damming certain Siberian rivers that normally flow into the Arctic Ocean, and diverting this fresh water to irrigate the farmlands. Climatologists fear such a project, for the damming of these rivers would reduce the fresh water flowing into the Arctic, which in turn would increase its salinity and promote the melting of its sea ice.

Although irrigation of an area leads to cool, wet weather, irrigation on a global level contributes to warmer temperatures. About 75 to 90% of irrigation water evaporates into the air, raising its water vapor content, producing cloudiness, and increasing the amount of precipitation in the region. This process results in cool, wet weather. On a global level, however, the change in surface albedo raises the average temperature level. Because it promotes the growth of vegetation, irrigation alters the balance between incoming solar radiation and outgoing terrestrial radiation; the decreased reflectivity of the surface area increases the net radiation gain. This in turn increases the global temperature level. As he concluded in his 1971 book, *Climate and Life*, Mikhail Budyko, Director of the Voeikov Geophysical Observatory in Leningrad, calculates this rise to be 0.07° Celsius, based on present irrigation rates.

By damming rivers and creating lakes, we again affect the climate and alter the radiational balance of the region. As they have a comparatively low reflective quality or surface albedo, open waters absorb incoming solar radiation and store the energy like heat reservoirs. This influence moderates the range of temperature extremes in the area. The region also experiences higher relative humidity and increased rainfall.

Even the chemical fertilizers used in agriculture can have severe repercussions on climate conditions, earth, and men and women. As fertilizers contain nitrogen, their use injects additional nitrogen into the atmosphere. In Chapter 8 we noted that solar flares can interact with atmospheric nitrogen and form nitric oxide. Since one nitrogen oxide molecule can react with 100,000 ozone molecules, even a small increase of nitrogen in the atmosphere can contribute to the depletion of the stratospheric ozone layer. Ozone, as we shall discuss later in this chapter, is the atmospheric constituent that blocks out the ultraviolet rays of solar radiation and shields the earth from potentially lethal radiation. Excessive use of fertilizers can lead to another case of overkill, whereby for the sake of maximal agricultural productivity, man so stimulates the soil as to destroy it. As above, so below. The introduction of chemical wastes and by-products into the environment produces injurious effects to both the atmosphere and the land.

Improper agricultural practices are the results of ignorance, greed, and the desire for immediate gratification. Their long-term consequences could prove cataclysmic. If we continue to degrade our environment, we can expect agricultural productivity to decline, land areas to become lifeless, climate patterns to change. Our inadvertant influences on the climate and the earth are not, however, limited to our land-use activities. Earth and climate changes are one of the major by-products of the industrial age in which we live.

Throughout their history, men and women have clustered into groups and built communities. To promote the growth and expansion of urban population centers, modern man takes undeveloped land, farm land, and forested areas and converts them into citylike environments of asphalt-paved roads, cement-constructed buildings, and large-scale industrial plants. During the eight-year period between the land-use surveys conducted by the U.S. Department of Agriculture in 1967 and 1975, about 2.5 million hectares of agricultural land in the United States were converted to spreading urbanization. Between 1960 and 1970, urban encroachment led to an annual loss of 0.25% of the agricultural land in West Germany and annual losses of 0.18% in both France and Great Britain. Such land conversions not only deplete the resources of land available for cultivation and food production, but they also change local weather patterns.

Commonly referred to as "heat islands," urban centers tend to be warmer than the surrounding rural areas. Cities, since they have a lower surface albedo than the countryside, affect the radiational balance. While the stone, concrete, and asphalt used to construct cities absorb and store heat, the vegetation of rural areas reflects more of the incoming solar radiation. Temperatures average about 1 to 2° Celsius higher in the city, but can soar up to 10° Celsius higher than surrounding rural areas. These temperature differences reach their extreme after sunset. As night falls, the rural areas cool rapidly but the cities, which release more slowly the heat stored in buildings and pavements, stay warm.

The radiational balance is affected not only by variations in surface features and their reflectivities. At the street level, cities experience a decrease in incoming solar radiation due to the shading from high-rise buildings, the increased cloudiness associated with cities, and the aerosol veil produced by the injection of particulate matter into the atmosphere from industrial sources, vehicular traffic, and other combustion processes. As we have noted, aerosol veils and clouds block out some of the incoming solar radiation and reflect it back towards space. Although incoming solar radiation reaching the city's surface is cut down to some degree, other factors make up for this decrease and lead to the temperature increases of urban areas. The lower surface albedo of the city absorbs more of the incoming radiation and reflects less of it back toward space. The increased aerosol veil and cloud cover, while blocking some of the incoming radiation, also prevents some of the outgoing radiation from escaping, and instead reradiates it back toward earth. This situation is particularly significant for urban areas, where a great amount of artificial heat is injected into the atmosphere. The emission of artificial heat comes from industrial plants, the exhausts of vehicular traffic, the winter heating of buildings, the summer use of air conditioning, and various other combustion processes. Artificial heat in cities can amount to as much as 10 to 15% of the solar energy received in various U.S. cities and is estimated to be as high as 33% in certain European cities. Consequently, urban centers are like heat domes with warmer temperatures than the surrounding countryside.

The aerodynamic roughness of the city's surface features significantly affects wind speed. While city streets channel the winds, high-rise buildings brake them. Wind speeds in cities are reduced by 20 to 30%. The varying heights of city buildings intensify the turbulence of air above the city. Added to these effects, the emission of heat from the city generates convection of hot air rising in plumes of updrafts. From this convection and the atmospheric input of water vapor from combustion, cloud formation over the city accelerates, and rainfall increases. Urban areas generally experience an annual precipitation increase of about 10% on an annual basis, although some cities have reported increased rainfall of up to 30%. This precipitation increase in urban areas is most apparent Mondays through Fridays, when the industrial activities of the city are in full gear. However, despite increased urban rainfall, the impermeability of the city surface causes the rains to run off rapidly, which in turn contributes to the lower relative humidity of cities.

The increased precipitation in cities influences regions

downwind of the city center as well, as has been evidenced by the documentation of Stanley A. Changnon, Jr. of the Illinois State Water Survey. Analyzing the records of precipitation, hail, and thunderstorms for the La Porte, Indiana, weather station, Changnon found significantly higher figures recorded than for other weather stations nearby. The fact that the figures showed an increase during recent years also intrigued Changnon. He noted a precipitation increase for the La Porte weather station of between 30 to 40% since 1925. Between 1951 and 1965, La Porte experienced 31% more precipitation, 38% more thunderstorm activity, and 246% more days with hail than did nearby weather stations. But why, Changnon wondered. He discovered a correlation between the precipitation increases in La Porte and the higher production levels of the iron and steel foundries in Chicago and Gary, Indiana, about 50 kilometers upwind of La Porte. In 1968, Changnon presented his conclusions to explain the increases. The industrial effluents emitted heat, moisture, and particles into the atmosphere upwind, stimulating cloud formation. The westerly winds then carried the clouds over Lake Michigan, where they picked up more water vapor. Over La Porte the clouds precipitated. From the data of the La Porte Weather Anomaly, Changnon was able to describe urban weather effects on downwind regions.

To test further urban weather effects, meteorologists have set up the Metropolitan Meteorological Experiment (METRO-MEX) in the St. Louis, Missouri, area. In 1971, in a 9850 square kilometer area around St. Louis, scientists began to install a network of some 250 sites where instruments measure and record the amounts of precipitation and other weather conditions. The weather is affected downwind of St. Louis, as far away as 40 kilometers east. With maximal intensity at between 16 and 24 kilometers from the city center, the downwind regions have experienced large increases in the frequency of rainfall, thunderstorms, and hailstorms. As the frequencies are fewer during the weekends, the increased precipitation is attributed to the city's industrial activity.

The effects of urban heat islands go beyond the immediate weather conditions we have been discussing. The injection of trace gases and particulate matter into the atmosphere may also wreak havoc with more general and long-lasting climate conditions. The combustion of fossil fuels such as coal, gas, and oil have continually increased the carbon dioxide concentration in the atmosphere. But not until 1938, when G.S. Callendar presented startling evidence of this gradual increase, did scientists begin to pay serious attention to the problem. Since 1850, the carbon dioxide content in the atmosphere has increased some 15%, from slightly less than 290 parts per million to approximately 335 parts per million. Estimates of the carbon dioxide concentration by the end of the century range from 365 to 400 parts per million. Of course, the estimated increases are dependent upon the output of carbon by the two main sources of emission, combustion of fossil fuels and deforestation. We have already noted the worldwide rapid rate of deforestation. When considering the global energy use of combustion fuels, we must anticipate a possible dramatic increase due to the industrialization of the developing nations. We must also accept the likelihood of governments' refusal to seek viable alternate energy sources, and the inherent risks of nuclear power. Some calculations estimate an increase of three to five times the present rate in the global energy use by the middle of the twenty-first century. Should this increase occur, the concentration of carbon dioxide might reach as high as 650 parts per million by the year 2050.

Not all of the carbon dioxide emitted from industrial sources or deforestation finds its way into the atmosphere. Some of the increased carbon dioxide concentration stimulates the process of photosynthesis and becomes absorbed by vegetation. But in our discussion of deforestation, we noted that vegetation, until recently considered a sink or absorber of carbon dioxide, is now believed to be a new source of it, due to intensive deforestation and land use changes. The only other recognized sink for carbon dioxide is the oceans. Oceans consist of two basic layers—a surface layer and the "deep sea" layer beneath it. The ability of oceans to absorb carbon dioxide depends largely upon their surface water temperatures. The lower that temperature, the greater the capacity to absorb carbon dioxide. Consequently, should the oceans become warmer, they would become less effective in absorbing, and might eventually become a *source* of carbon dioxide. According to various calculations, the oceans presently absorb less than 50% of the carbon dioxide produced. Minze Stuiver of the University of Washington estimates that 34% is absorbed by the surface

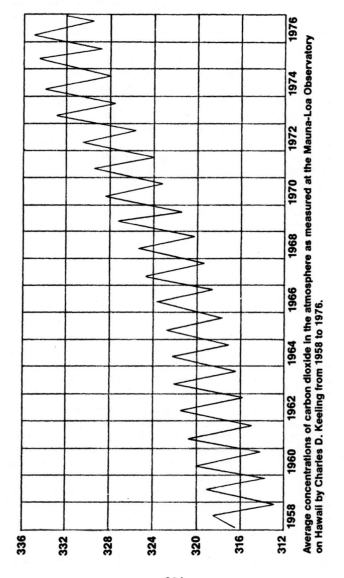

Average concentrations of carbon dioxide in the atmosphere as measured at the Mauna-Loa Observatory on Hawaii by Charles D. Keeling from 1958 to 1976.

214

layer of the oceans and another 13% by the "deep sea." Thus over 50% of the carbon dioxide generated by man's activities remains in the atmosphere, creating a "greenhouse effect" that also affects the earth's radiational balance.

As we mentioned in discussing deforestation, the "greenhouse effect" results from the effect of carbon dioxide on outgoing terrestrial radiation. While atmospheric carbon dioxide does not affect the incoming, short-wave solar radiation, it does block some of the outgoing, long-wave terrestrial radiation, absorbing some of it and reradiating the rest back to the earth. This disturbance of the radiational balance promotes a warming trend. Based on the climate model designed by Syukuro Manabe and Richard Wetherald, a doubling of the atmospheric content of carbon dioxide might lead to an increase of 2.4° Celsius in the global mean temperature. The warming would be more pronounced at the higher latitudes toward the polar regions and less pronounced at the lower latitudes toward the equator. In their model, Manabe and Wetherald assume fixed conditions of relative humidity and cloudiness. In nature, however, this variable would not be fixed. Since one effect of higher atmospheric concentrations of carbon dioxide is likely to be increased rainfall, we would expect cloudiness to increase. But increased cloud cover would have the opposite effect to increased carbon dioxide content. As it would block some of the incoming solar radiation, increased cloud cover would promote cooler temperatures. An increase of around 2.5% in the global cloud cover could offset a doubling in the concentration of carbon dioxide. In additon, Stephen Schneider, of the National Center for Atmospheric Research in Boulder, Colorado, and S.I. Rasool, of the National Aeronautics and Space Administration, suggest that the rate of temperature rise from atmospheric carbon dioxide would diminish as greater quantities of the gas were added to the atmosphere.

While increases of carbon dioxide in the atmosphere may prompt a net warming effect, the injection of particulate matter may do just the opposite. As we have seen, carbon dioxide allows short-wave, incoming radiation access to the earth's surface, but reduces the amount of long-wave, outgoing radiation that can escape the earth systems. In contrast, particulate matter is similar to cloud cover; it prevents some of the incoming solar radiation from reaching the earth's surface. There-

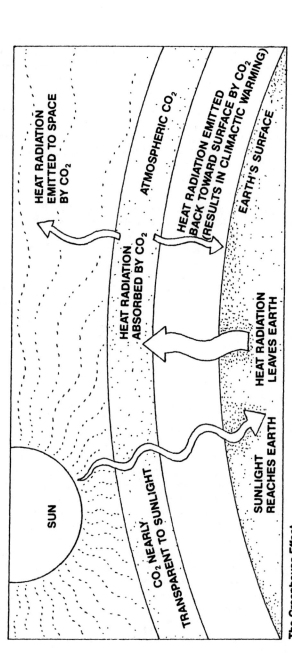

The Greenhouse Effect.

fore, an aerosol veil would diminish the incoming radiation, whereas increased carbon dioxide would reradiate the outgoing radiation back toward the earth's surface. Will these two countervailing influences cancel each other out? Will the carbon dioxide increases dominate, triggering a warming phase? Or will the aerosol effect prevail, contributing to the natural cooling trend? The answer to these questions will help to determine man's influence on the climate and future climate conditions on earth. But before we take up these questions, we must first consider man's effects on another atmospheric constituent, the ozone layer in the stratosphere.

Several times in this book we have underlined the importance of ozone, which absorbs the ultraviolet rays from the sun, preventing this potentially lethal radiation from reaching the earth. In Chapter VIII, we mentioned Paul Crutzen's theory that ozone is depleted when it reacts with the nitric oxides formed from the dissociation of nitrogen by electrically-charged particles emitted by solar flares. Earlier in this chapter, we cited agricultural fertilizers, with their nitrogen base, as potential destroyers of stratospheric ozone. One molecule of an oxide of nitrogen can react with as many as 100,000 ozone molecules. However, oxides of nitrogen are not the sole sources of stratospheric ozone destruction.

In June 1974, F. Sherwood Roland and Mario Molina of the University of California, Irvine, proposed a theory of ozone depletion that condemned many of the household products we use on a daily basis. Roland and Molina claimed that aerosol propellants employed in the spray applicators of everything from perfumes to dessert toppings, shaving creams, deodorants, and other items posed a threat to the stratospheric ozone. Aerosol propellants and refrigerants are composed of chlorofluorocarbons, fluorocarbon-11 (trichlorofluoromethane, CCl_3F) and fluorocarbon-12 (dichlorodifluoromethane, CCl_2F_2). According to the Roland-Molina theory, the fluorocarbons, instead of being washed out in the troposphere, remain inert and float up into the stratosphere, where they are broken down by the process of ultraviolet photodissociation and release chlorine molecules. Like a nitrogen oxide molecule, the chlorine molecule, through catalytic chain reactions, can destroy thousands of ozone molecules, and lead to the depletion of the ozone layer of the stratosphere. The Roland-Molina theory gained acceptance in the United States, where on March 15, 1978, the government announced a ban, effective December 15, 1978, on the manufacture of virtually all aerosol propellant products that use chlorofluorocarbons.

While this government ban is to be applauded, another government regulation has added to the problem of ozone depletion. Through the efforts of the Environmental Protection Agency, the U.S. government has determined that the use of the solvent trichloroethylene, employed in the degreasing and cleaning of metals, should be discontinued, and a substitute of the compound methyl chloroform (CH_3CCl_3) used. Although this decision was based upon the fact that methyl chloroform is relatively inert in the troposphere compared to trichloroethylene, it did not take into account the fact that methyl chloroform's tropospheric life span is about 10 years, as compared to trichloroethylene's life span of a few days. Methyl chloroform is also used as a dry cleaning solvent. Calculations show that as much as 15% of the methyl chloroform released into the atmosphere makes its way into the stratosphere. Once there, it releases chlorine molecules, as it is broken down by ultraviolet photodissociation. Like the fluorocarbons, these chlorine molecules can destroy the stratospheric ozone through catalytic chain reactions. Estimates indicate that the atmospheric level of methyl chloroform is increasing at a rate of about 20% a year: thus this solvent could prove of serious consequence to ozone depletion.

Even if all ozone-destroying compounds were banned, the process of stratospheric ozone depletion would continue for many years into the future. Methyl chloroform has an estimated average atmospheric life span of 10 years, fluorocarbon-11 an average of 45 years, and fluorocarbon-12 an atmospheric life span of 68 years. A comprehensive ban, however, seems unlikely at the present. Despite the U.S. government's forbidding the use of chlorofluorocarbons in aerosol propellants, these products account for only 25% of the world's chlorofluorocarbon use. Although there are complex natural processes by which ozone is replenished in its 11-year cyclical variations, calculations of man-made ozone destruction show the possibility of an eventual depletion of stratospheric ozone in the range of 2 to 20%. A 15% reduction in ozone could mean a 30% increase in the amount of ultraviolet radiation reaching the earth's surface.

Increased exposure to ultraviolet radiation would harmfully affect life on earth. Acting on the reproductive molecule, DNA, and protein synthesis, ultraviolet radiation can thwart the growth of vegetation and promote genetic mutations in various species of life-forms. During times of polarity reversals in the earth's magnetic field, which coincided with intense depletion of stratospheric ozone, the earth has been showered with lethal radiation, causing genetic mutations and even elimination of some species of earth life-forms, as well as the creation of others. Even minor changes in ultraviolet radiation have increased the incidence of skin cancers among humans. Theoretically, a 1% depletion of stratospheric ozone can produce a 2% rise in skin cancers. Man-made destruction of the stratospheric ozone, combined with the ozone depletion arising from catalytic chain reactions triggered by solar flares, can turn the periods of maximal sunspot activity into times of high risk of skin cancers and to subtle genetic mutations.

Ozone depletion also influences the earth's climate, for ozone destruction alters the radiational balance. A 15% reduction in stratospheric ozone could produce a temperature rise of 10° Celsius in the upper stratosphere, while the consequences for the troposphere and earth's weather conditions are subject to debate at present. What is certain is that ozone reduction will cause significant weather changes.

When, in the early 1970's, the United States was debating the development of a supersonic transport (SST) airplane, a major question raised was the effects of the effluents from the SST's engines on the atmosphere and, more specifically, on the stratospheric ozone. The proposed American SST, the Boeing 2707, was projected to fly at an altitude of about 20 kilometers, in contrast to the altitude of 17 kilometers flown by the Russian-built SST, the Tupolev 144 and the British-French SST, the Concorde. The Boeing 2707 was also projected to have three times the fuel flow of these smaller SST's. Opponents of development claimed that the Boeing SST would seriously aggravate an already worrisome situation. Since the SST's fly at altitudes where their exhaust products can have atmospheric lifetimes averaging one to two years, their nitrogen oxide wastes would accelerate the destruction of stratospheric ozone. According to models calculating the effects of a fleet of 500 SST aircraft, several scientists have estimated a potential depletion

219

of between 2% and 25% in stratospheric ozone by the year 2000. The higher figure was based on a fleet of American-type SST's, which would have flown at higher altitudes and would have had the greater fuel flow. As we know, the American SST was shelved. The Concorde survives as an operational SST aircraft, but its wastes, which contribute to the depletion of the stratospheric ozone, are not negligible.

No form of air travel, however, is without its harmful effects. While SST's may affect the stratospheric ozone, jet aircraft flying at lower altitudes emit sulfur particles and leave condensation trails in the atmosphere. This increased particulate matter adds to or promotes an aerosol veil, reflecting back some of the incoming solar radiation. As they tend to form artificial clouds, the condensation trails, heavy with water vapor, may drop ice crystals on lower clouds, in turn nucleating them and triggering precipitation.

It is the power of nuclear energy which has raised the stakes exponentially in the game industrial society plays with the natural forces around it. Although we can take pride in our knowledge of the laws of energy and nature on which this nuclear power ''move'' is founded, we should recognize that we have, by this same move, placed ourselves in a position of extreme jeopardy. We have already noted the eerie propensity on the part of United States authorities to select nuclear reactor sites that are dangerously close to geological fault lines. There is also a frightening lack of awareness concerning the by-products and wastes of nuclear energy which are returning to plague our environment. The gaseous wastes of nuclear energy are altering the earth's atmosphere.

Atmospheric testing of nuclear weapons injects oxides of nitrogen into the air which, through catalytic chain reactions, destroy the ozone in the stratosphere. It is alleged that the atmospheric testing of nuclear weapons in the early 1960's has already depleted the stratospheric ozone by several percent. It is possible that the distortion of the earth's magnetic field, which also results from the testing of nuclear weapons, will reduce the resistance between the earth and the ionosphere. If this were to happen, the earth's protection from the lethal radiation of solar and cosmic rays would be considerably lessened, and we would be subject to the toxic effects of radioactivity.

Fossil-fuel generating plants are about 40% efficient, which means that 60% of the heat they generate is released into the environment as waste. The present generation of nuclear reactors is only 30% efficient; thus, more than approximately 70% of the heat generated is released as waste. Some of this waste is released into the atmosphere as the radioactive gas krypton-85. Not only does krypton-85 expose the environment to traces of radioactivity, but it also increases the conductivity of the lower atmosphere, thereby reducing the resistance between the earth and the ionosphere.

If we continue to flout the dangers inherent in the use of nuclear energy, the consequences will no doubt be devastating. The range of horrific effects, from radioactive poisoning to cancers and other genetic mutations, which, as we have shown, are associated with ozone depletion, will make the radical weather changes seem to us secondary. The decision to abandon nuclear energy and to persevere in developing less toxic energy sources will be a difficult one. To make it, we must overcome the resistance of many vested interests. Not to make it, however, is eventually to bow to forces much more powerful than the mere forces of human resistance.

In this chapter we have seen how man's actions contribute to the aerosol veil. Recent estimate sindicate that man's activities produce between 5% and 45% of all particulate matter in the atmosphere. Aircraft exhausts, industrial effluents, photochemical reactions of emitted gases, desertification, overgrazing, deforestation, "slash and burn" practices, overplowing, and ill-managed irrigation schemes—all inject particulate matter into the atmosphere. Add to these the day-to-day natural sources of particulate matter such as ocean spray, smoke from forest fires, wind-blown dust, pollen, spores, and microrganisms of living matter; include the particles injected by the geophysical variations in volcanic eruptions and earthquake activity, and we have a fairly adequate enumeration of the sources which impel us toward a cooling trend. For they all contribute to the production of an aerosol veil which blocks out some of the incoming solar radiation and promotes cooler temperatures and cloud formations which have the cyclical effect of blocking out even more solar radiation.

Sometimes the optical properties of the particles or the optical depth of the veil are such that they have the opposite effect of allowing solar radiation to pass through it while blocking outgoing terrestrial radiation, thus causing a warming trend. An aerosol veil over areas of higher surface reflectivity, such as the polar regions, would similarly produce this opposite effect. The warmer local temperatures in the polar regions slow atmospheric circulation and reduce heat transport from the equatorial latitudes. Consequently, the net effect of aerosol veils would still be a cooling trend in the polar regions, for there the warmer local temperatures would still be lower than the reduced heat flow from the lower latitudes. Thus, while local variations may produce anomalies of warmer temperatures, the global effect of the aerosol veil remains the same—the instigation of a cooling trend.

From calculations made by Stephen Schneider and S.I. Rasool, we can predict that an increase in the global aerosol background concentration by a factor of four could reduce the mean global temperature level by as much as 3.5° Celsius. During the 60 years between 1910 and 1970, the atmospheric content of particulate matter is estimated to have doubled. More recently, the injection rate of particulate matter into the atmosphere has accelerated. Largely through human activity, the rate of atmospheric particle increase is seen as doubling again in the next 10 to 20 years. While increases in the atmospheric carbon dioxide concentration will increase, and continue to promote a warming trend, evidence indicates that increases in particulate matter will be greater, and will more than offset the effects of carbon dioxide concentration. The actions of men and women can thus be seen as having an aggravating and deciding influence on the global cooling trend. If this hypothesis is valid, we can expect the cooling to become especially evident after the present cycle of sunspot activity peaks around 1982.

In the following chapter, we shall see how our potential for effecting radical climate changes in the future has been actually acted out and thus prefigured in the recent past. It is hoped this clear-sighted recognition will act as a curb to our compulsion to repeat.

XII

WEATHER MODIFICATION AND WEATHER WAR: MAN PULLS THE TRIGGER

Consider this joke from Freud's book on jokes: L., who has borrowed a pot from R., returns it, and is then sued by R. because the pot now has a big hole in it which makes it useless. L. defends himself by saying that first, he has never borrowed the pot; second, the pot already had a hole in it when he received it; and third, that he returned the pot undamaged. This is certainly not a very funny joke, but it makes a valuable point about the absurdity of human defensiveness. We can understand this point more clearly if we imagine the pot to be the environment, the borrowed home of man, and the three responses the various denials man makes regarding his relationship to his environment. In the first case he simply refuses to acknowledge the existence of this environment; in the second, he admits that it is there, but denies his own ability to affect it; in the third, he tacitly admits that he can affect the environment, but claims innocence. Often, absurdly enough, these denials are maintained simultaneously. It will be the purpose of this chapter to demonstrate the illogic of this defense; to show that human activity is, in fact, guilty of contributing to major climatic fluctuations; and to suggest the penalties, the retributions which may be extracted by R., the forces of nature, for our transgressions.

In the early 1970's the world became aware of a grim reality—the monsoons had failed the Sahel south of the Sahara, and drought was taking its toll of the populace. From their television sets, the comfortably situated developed nations stared at the hollow eyes of children bloated with famine and disease. The anguish of the Sahel was apparent. It was a reality largely confined to the southern edge of the Sahara, but one that threatened the lives of upward of six million people and made refugees of an estimated 10 million, virtually half the population of the west African region.

Composed of the six sub-Saharan countries of Mauritania, Senegal, Mali, Upper Volta, Niger, and Chad; the Sahel had suffered severe droughts before; during 1900 to 1903, 1911 to 1914, and in 1930 to 1931. But the toll was never as heavy as that exacted between 1968 and 1974. This drought finally ended in 1974, but we should ask ourselves: why was this one such a catastrophe?

Almost half the world's population is dependent on the monsoon. The word itself comes from the Arabic "monsoon" which means "season," and describes the seasonal winds that reverse direction. Along with this reversal in wind direction come changes in other climatic patterns, such as temperature, water vapor, cloudiness, and precipitation. The monsoons are often associated with rainfall, for it is at the time of the monsoon that areas of Australia, southeast Asia, India, and northern Africa experience their annual rainfall. In the models designed by Syukuro Manabe of Princeton University and his colleague Jigadish Shukla, monsoons are seen as dependent on the surface temperatures of the oceans. As the land heats up during the growing season, so do the oceans, which absorb more of the sun's warmth. The heating of the land provokes the convection of hot air, and the rising hot air forms the low-pressure systems described in Chapter IX. These low-pressure systems draw moisture from the oceans. Manabe and Shukla found from their monsoon models that the warmer the sea surface, the more moisture is absorbed, whereas the cooler the sea surface, the less moisture is absorbed. In the late 1960's and early 1970's, the natural global cooling trend resulted in less rainfall potential from the monsoons. Prompted by the cooling trend, changes

224

in atmospheric circulation patterns aggravated the situation by forcing the monsoon rains that should have fallen on the Sahelian countries further south.

One of the ironies of the 1968 to 1974 Sahel drought is that, had modern practices never touched the lives of the Sahelian populations, they might have been better able to cope. Once a nomadic culture, the peoples of this arid region were traditionally adapted to the cyclical periods of drought. When drought struck the region, the livestock would graze the permanent watering areas. During wet periods the livestock would graze further afield, allowing the vegetation around the watering areas a chance for regrowth. Like the rest of us, the people of the Sahel fell out of synch with the natural cycles. Improvements in health practices produced a dramatic increase in the population. To feed this population, larger herds of livestock were needed. To alleviate the water shortage, wells were drilled, thereby allowing for more livestock and greater cultivation of the land. Overgrazing and overcultivation, led to a loss of vegetation. The harvesting of firewood for fuel stripped the land, and opened it up to the potential ravages of a recurring drought. The people had forgotten the natural cycles of their own environment. When the monsoons failed in 1968 and in the years following, the inhabitants were ill-prepared.

The first reaction to the drought was to drill more wells to maintain the livestock. But this only accelerated the denuding of the land. Desertlike conditions swept across the Sahel. Extensive crop failures occurred, and about one-third of the livestock perished. Hunger preyed upon the people. Hundreds of thousands are estimated to have died not only from famine but also from the associated diseases of gastroenteritis, cholera, and pneumonia. Until the developed nations came to their rescue, it was feared that as many as six million people might die from the Sahel drought. Although much of the relief food came from the United States, we will conjecture later that U.S. war operations in southeast Asia indirectly contributed to the drought in the Sahel.

While the agony of the Sahel was perhaps the most extreme situation, other areas were severely affected by the adverse weather conditions of this period. To the east of the Sahel, drought and famine plagued the Sudan, Ethiopia, Somalia, Kenya, and Tanzania. In October 1973, the government of

225

Ethiopia appealed for aid, acknowledging the deaths of 100,000 people and a continuing death rate of 2000 to 3000 people each week from famine. When the passage of time allows for a more dispassionate perspective on recent climatic history, 1972 will stand out as a year of dramatic weather change and catastrophe.

In India, the monsoon came three weeks late, causing a loss in food production of about 33%. The harvests in parts of India declined by as much as 60%, while in Bangladesh the rice harvest fell 2.5 million tons below expectation. The drought in the U.S.S.R. produced a wheat harvest 12% below expectation and necessitated large purchases of wheat from the United States. China, Australia, and South America also endured drought conditions. In Peru, the change in ocean currents disrupted the fishing industry. Although in 1971 it enjoyed the largest fish catch in the world, Peru's catch of anchovies in 1972 was cut by more than one-half, forcing the closing of many fisheries. While the grain-growing regions of North America were drenched with too much rain, Europe encountered abnormally high temperatures and parched conditions. The year 1972 was one of abnormal weather throughout the world. It claimed the lives of several hundreds of thousands and threatened the existence of millions.

Although modern technology, the development of high-yield seed varieties, and the optimal weather conditions of the twentieth century prompted the "Green Revolution," and promised the eradication of hunger in the world, man must now confront a decline in food production. Despite recent increases in food production, the rates of increase have been declining. In the 1950's, food production rose at a rate of 3.1% a year, in the 1960's by 2.8%, and in the 1970's by 2.4%. In 1972, when the world endured abnormal weather, food production actually fell by 1%. Current increases in food production are not keeping up with the needs of the world population. In 1970, an estimated 400 million people were suffering from chronic malnutrition. By 1978, that figure had jumped to 455 million. During the 25 years between 1975 and 2000, food demand is expected to double. And the future prospects for food production are alarming.

According to Lester Brown of the Worldwatch Institute, the cereal yield per hectare between 1950 and 1971 rose from 1.14 metric tons to 1.88 metric tons per hectare, an annual incease of 2.4%. Between 1971 and 1977, the increase in cereal yield per hectare increased at an annual rate of only 0.6%. The land can only give so much of itself before it can give no more. No matter how we seek to extort increased agricultural output per hectare, our rape of the land cannot continue unabated without our exhausting it. Soil fertility is already on the decline. Recent soil surveys show a drop in inherent fertility on 20% of the world's cropland. Combined with desertification, strip mining, construction of dams and highways, encroaching urbanization, and other changes in land use, soil degradation and erosion are shrinking the land suitable for cultivation. Brown estimates that the potential availability of future croplands has an upward limit of only a 10% increase in the future. Other agronomists calculate that the loss of prime cropland may approach 50% of its present area by the turn of the century.

To add to these gloomy prospects, we must realize that a global cooling trend can only aggravate the situation. During a cooling trend, worldwide food production would fall sharply. In the higher latitudes toward the polar regions, expanding snow cover would make the cultivation of land impossible. In the higher middle latitudes, the annual growing season would shorten, and the frequency of unseasonal "killing frosts" would increase. Stagnant weather patterns associated with a cooling phase would lead to regional extremes, with some areas experiencing extensive flooding and others suffering severe drought. Drought, in fact, would pose the greatest threat to the world during such a cooling period. As in the early 1970's, the monsoons would fail. Again, the people of the Sahel would confront the specter of death. Any rains that might come would fall further to the south. The peoples of the world would very likely flail out in agony as hunger ripped away the semblance of community, leading to undignified death. Within 25 years of a pronounced cooling trend, starvation might claim the lives of as many as 500 million.

The bountiful harvests of the late 1970's have allowed us to become complacent. The tragedy of 1972 is vaguely remembered as but a brief nightmare, a passing challenge for survival rather than the omen that it was. We have arrogantly elevated ourselves to a position of false confidence, believing that nothing is beyond our understanding or control. Our insecure need for continual expansion and exploitation have left an indelible mark on the human psyche. Today we live in a dream world of blind hope and omnipotent virility. As a result, we have cast ourselves adrift from the reality of life, the reality of nature's alternating cycles. We scoff at warnings from those whom we characterize as "prophets of doom." And yet it would be wise to recall the story of Joseph and the Pharaoh as recorded in Genesis.

Even as a child, Joseph was subject to prophetic dreams. Jealous of his favored status with their father, his brothers conspired against Joseph and sold him to the Ishmaelites, who took him to Egypt. While imprisoned in Egypt, Joseph interpreted the dreams of two servants of the Pharaoh, and the dream images came to pass as Joseph had interpreted them. Two years later, the Pharaoh was sorely troubled by a dream. He called for all the magicians and wise men of Egypt to interpret his dream. But none could do so. One of the Pharaoh's servants mentioned to the Pharaoh the accuracy of Joseph's interpretations. Joseph was summoned. When he had heard the Pharaoh's dream, Joseph interpreted it in the following manner:

"Behold, there shall come seven years of great plenty throughout all the land of Egypt:

And there shall arise after them seven years of famine; and all the plenty shall be forgotten in the land of Egypt; and the famine shall consume the land;

And the plenty shall not be known in the land by reason of that famine following; for it shall be very grievous. . . .

Now therefore let Pharaoh look out a man discreet and wise, and set him over the land of Egypt.

Let Pharaoh do this, and let him appoint officers over the land, and take up the fifth part of the land of Egypt in the seven plenteous years.

And let them gather all the food of those good years that come, and lay up corn under the hand of Pharaoh, and let them keep food in the cities.

And that food shall be for store to the land against the seven years of famine, which shall be in the land of Egypt; that the land perish not through the famine."

—Genesis XLI:29-31, 33-36

Pharaoh heard the truth in the interpretation and appointed Joseph as overseer for the storing of grain reserves. Seven bountiful years of harvests came, after which the years of dearth began, as foretold. Although hunger plagued the peoples of the earth, Egypt had prepared for the adversity and was spared the agony suffered by other nations, for Joseph opened the storehouses and distributed the food reserves to the people.

In the twentieth century, the world's population must again prepare for significant shortfalls in harvests and food production due to adverse changes in climate conditions. Food reserves must be built up. Evidence indicates a natural cooling trend. Geophysical variations, magnified by man's actions, will intensify the cooling trend during the 1980's. The time to prepare for this is now. Calamity may be upon us all too soon.

Most of us are prone to put off hard decisions, opting instead for the satisfactions of the moment. If the warnings go unheeded, if preparations are neglected, it might be too late when it becomes apparent that we are locked into a struggle for survival. Technology in itself is neither good nor evil. The uses it is put to will determine the outcome for humanity.

Man has already used technology to modify weather patterns and control the natural forces. Through his knowledge of the formation of raindrops, he has explored the possibilities of triggering their artificial formation. Known as cloud-seeding, this technique was first developed in 1946, through research done at General Electric in Schenectady, New York, by Vincent Schaefer, Bernard Vonnegut, and Irving Langmuir. Here's how cloud-seeding was developed: through the convection process, warm, moist air rises and cools in the lower temperatures of the higher altitudes. At those lower temperatures, the air cannot hold as much water vapor. When the dew point is reached, the water vapor condenses on the particulate matter in the atmosphere. At the stage of condensation, heat is released and raises the air higher into the atmosphere. The air cools further. When the temperature falls below freezing, the moisture content of the air solidifies on the particulate matter and forms ice particles. As the ice crystals grow large and heavy, they fall and reach the earth as the precipitation of snow, hail, or rain. Drawing on the need of particulate matter to serve as the nuclei on which precipitation can form, Schaefer proposed substituting man-injected particles to serve as the nuclei.

Using a home freezer unit to carry out the experiments, Schaefer and Langmuir found that dry ice (solid carbon dioxide) would generate the formation of ice crystals in clouds. On November 13, 1946, Schaefer tested the practical application of his concept of cloud-seeding. From a plane he dropped several pounds of crushed dry ice into a cloud over Mount Greylock in western Massachusetts. Within minutes, the water droplets in the cloud had changed into ice crystals. Man's potential to trigger precipitation became reality.

Shortly after Schaefer's successful tests, Vonnegut suggested the seeding of silver iodide, since its atomic structure closely resembles that of ice crystals. At the present time, silver iodide is the most commonly used compound in cloud-seedings; lead iodide, liquid propane, and even table salt have also been successfully used to seed clouds. The seeding compound is introduced by one of several means: dropped from an airplane, fired into the clouds from anti-aircraft guns, or heated to form smoke, which is carried aloft into the clouds. Studies show that cloud seeding can diminish as well as increase precipitation.

From his cloud-seeding experiments in the Rockies near Climax, Colorado, during the 1960's, Lewis O. Grant of Colorado State University, established that seeding would increase snowfall when the temperature of the cloud top was $-26°$ Celsius or warmer. When cloud-top temperature was colder than $-26°$ Celsius, cloud-seeding would diminish the snowfall. This finding was confirmed by the cloud-seeding experiments carried out near Steamboat Springs, Colorado, by J. Owen Rhea and L. G. David of E.G.&G., Inc., of Boulder, and by the Bureau of Reclamation San Juan Pilot Project in southern Colorado. Seeding the colder cloud tops created an overseeded condition in which the nuclei of the ice crystals were so small that they evaporated instead of precipitating.

Practical application of cloud-seeding has primarily focused on hail-suppression projects. Hailstorms develop out of thunderstorms. Extending to heights of several kilometers, the thunderheads (turbulent cumulonimbus clouds) exhibit an awesome convective process with violent updrafts and downdrafts powering the vertical exchanges of warm and cool air. As warm, moist air is sucked up into the thunderheads at velocities reaching a maximum of 18 to 31 meters a second, the water droplets are lifted until they grow large enough to overcome

230

the force. If the maximal velocity of the updraft occurs at a cloud level where the temperatures are above − 5° Celsius, the precipitation will generally fall as rain. At cloud level temperatures of between − 5° Celsius and − 20° Celsius, precipitation will generally fall as large hailstones. At cloud-level temperatures below − 20° Celsius, precipitation will fall as many small hailstones or as snow. By seeding hailstorms, we increase the number of nuclei around which water droplets can condense, and thereby produce many smaller hailstones as opposed to fewer but larger hailstones. The smaller hailstones will either melt on their way to earth or they will cause less damage to crops when they fall.

The world's major hail belts are located in the plains of North America, the Argentine pampas, central Europe, northern India, the Russian Caucasus, and central Africa. Hail-suppression projects have been used in these areas, in such countries as the United States, Canada, Argentina, France, Germany, Italy, Kenya, and the Soviet Union. The Soviet Union claims that cloud-seeding has reduced crop damage from hail by as much as 80 to 90% when compared to preseeding records and to damage in nearby nonseeded areas. Believed to suffer the highest incidence of hail in the world, the Kericho region of Kenya experiences thunderstorms more than 200 days a year with 85% or more of these storms producing hail. As the hailstones severely damage the yields of the tea-growing industry in the region, a three-year hail-suppression project was instituted in 1967. The results for the first two years of the program showed the average tea loss was 2863 pounds per seeded storm, as compared to an average loss of 6894 pounds for nonseeded storms. In the United States, where crop losses from hail can run as high as $1 billion annually and usually claim more than 25% of the annual wheat crop, cloud-seeding experiments in South Dakota were found to reduce the energies of hailstorms by 60%. However, as we shall consider shortly, there are many unknowns involved when tampering with the weather, unknowns that can trigger catastrophic consequences and render the supposed cure worse than the disease.

Weather modifiers have even seeded hurricanes in an attempt to reduce their destructive intensity. Called cyclones in the Indian Ocean and Australia, typhoons in the western North Pacific, and hurricanes in the western Atlantic; these storms are all of one genus known as "tropical cyclones." They wreak havoc along coastal regions, as did the cyclone which struck Bangladesh in November 1970, killing an estimated 300,000 people.

Cyclones form in latitudes slightly beyond the equator, where the earth begins to slant inwards toward the polar regions, causing the spiraling of surfacewuinds both from the sideway deflection of the earth's rotation and from the transfer of heat from the equator to higher latitudes (cf. Chapter IX). Cyclones tend to form in the late summer or early autumn, when the surface sea temperatures have warmed to a level of at least 26° Celsius. Moving over the oceans, low-pressure centers draw up the warm, moist air, producing clouds and rain. The release of heat intensifies the low-pressure system. When extreme intensity builds up in the low-pressure system, it can turn into a cyclone. In the middle of the cyclone is its "eye," characterized by an eerie calm of soft winds, translucent clouds, and temperatures considerably higher than the exterior. Surrounding the eye is the "eyewall," where violent updrafts of air with velocities approaching 185 kilometers an hour (roughly 100 knots) power the storm to cyclone force. Creating a vacuumlike condition, the cyclone exhibits a marked drop in atmospheric pressure and consequently draws in even more air from the surrounding area, adding fuel to drive it.

The first attempt to seed a hurricane took place on October 13, 1947, when Irving Langmuir seeded a hurricane off the southeast coast of the United States with 200 pounds of crushed dry ice. The hurricane veered off course on an erratic path, bringing it inland. Langmuir believed that this had possibly been triggered by the seeding.

Under the auspices of the National Oceanic and Atmospheric Administration, the U.S. government instituted Project· Stormfury to study the feasibility of hurricane modification through cloud-seeding. Rough tests of seedings were conducted on Hurricanes Esther in 1961 and Beulah in 1963. The results indicated a reduction of 10% in wind speeds. The concept for the seeding of a hurricane emphasizes reducing the temperature differences between the interior (eye) of the storm and the storm's exterior (eyewall), thereby reducing the wind intensity and modifying the force of the hurricane. Consequently, the clouds on the outside edge of the eyewall were seeded. In 1969, Project Stormfury assaulted the fury of Hurricane Debbie.

On August 18, 1969, Hurricane Debbie had reached maximal wind speeds of 182 kilometers an hour at an altitude of 3659 meters. Five times, at two-hour intervals, the eyewall clouds were seeded. Five hours after the seedings, the maximal wind speeds were measured at 126 kilometers an hour, a drop of 31%. By August 20, Debbie had regained her strength. Maximal wind speeds were 183 kilometers an hour. Debbie was reseeded; six hours later, the maximal wind speeds measured 156 kilometers an hour, a drop of 15%.

In 1971, a new concept for seeding hurricanes was suggested. This idea proposed seeding the storm clouds away from the eyewall in order to promote the formation of a second eyewall. Surrounding the storm's inherent eyewall, but at a distance removed from the eye of the storm, this second eyewall was intended to take up some of the incoming flow of moist, warm air before it reached the inherent eyewall. By sapping the original eyewall of the amount of energy it would otherwise receive, this process would diminish the maximal wind speeds of the storms, and reduce the intensity of the hurricane.

Although experiments on seeding hurricanes do suggest a potential for reducing maximal wind speeds and thereby dissipating a storm's strength, there are definite risks involved in tampering with intense natural forces. As Langmuir noted in his initial tests, hurricane seeding may knock the storm off its path onto an erratic course with possible catastrophic consequences for inland regions. Although the initial seedings of Hurricane Debbie seemed to reduce its intensity, the storm regained even more than its initial strength. Could seedings sap storm strength momentarily, only to have it grow more forceful? Could seedings produce worse storm damage than would occur had the storm worked through its energy according to its natural dynamics? Our present understanding of hurricane dynamics and other natural systems is too incomplete to answer these questions. It's still too early to consider hurricane modification practicable.

Despite our incomplete understanding about the mechanics of weather, we continue our efforts to modify it. Although weather modification has been employed successfully in precipitating rainfall, reducing crop damage from hailstorms, and diminishing wind speeds of hurricanes, the overall results are questionable. Success has been mixed with failure. Certain attempts at weather modification may have wrought disastrous consequences. Other attempts have served malevolent purposes.

In June, 1972, Rapid City, South Dakota, was devastated by flooding. Subjected to cloud-seeding prior to unleashing its wrath on the small midwestern city, this storm claimed 250 lives and caused property damage calculated at $100 million. Whether the cloud-seeding contributed to the destructiveness of the storm has been debated among meteorologists. Some insist that the storm's intensity and the cloud-seeding were solely coincidence. If it were merely coincidence, then coincidence may again have played a part in the storm that struck southern California in February 1978. While rain was falling in some areas of Los Angeles County, the county's Flood Control Department was seeding clouds near Big Tujunga Canyon in an effort to increase the precipitation. Within hours of seeding, the storm intensified, becoming one of the worst ever to hit southern California. Big Tujunga Canyon was subsequently flooded. Property damage ran close to $50 million, and some 20 people were killed due to the flooding.

Although these two examples (and others like them) linking cloud-seeding and storm damage may indeed be coincidental, there is no doubt that weather modification can have adverse consequences. The impact of cloud-seeding is not confined to the local area where it is carried out. Studies have shown that cloud-seeding can have downwind effects at distances up to 1450 kilometers away. If man changes the weather in one area, he changes it in other areas as well. Consequently, the decision to seed clouds by one locale, or by one country, can impose effects of that decision on other locales or countries. While one area, parched from lack of rain, might feel compelled to seed any cloud offering the potential for precipitation, the premature triggering of precipitation could rob downwind areas of their expected rainfall. Even within the same locale, cloud-seeding operations might favor one group of interests to the detriment of another.

The possibility of regional, or even international, disputes because of cloud-seeding is not only likely but has already been the subject of controversy. One example was the cloud-seeding operation carried out in the San Luis Valley of Colorado during the mid-1960's and into the early 1970's. Initiated by the Adolph Coors Company to reduce hail damage to the Moravian barley crop, cloud-seeding was favored by the barley growers, but opposed by the more numerous ranchers who feared drought more than they feared hail. In a November 1972 referendum dealing with weather modification in the San Luis Valley, the five-county area voted four to one against its use.

The negative effects of weather modification have been consciously exploited for the malevolent purposes of military warfare. During the United States involvement in the war in Indo-China, weather modification was employed to intensify the monsoon rains and slow down enemy troop movements.

Allegations claim that the C.I.A. (Central Intelligence Agency) initiated weather modification in Vietnam with cloud-seeding operations over Saigon as early as 1963. In October 1966, a control test of seeding techniques was carried out over the Laos panhandle. From the 56 test seedings carried out, over 85% of the seeded clouds were triggered to precipitate. The full-fledged use of weather modification in the war began March 20, 1967, and continued until July 5, 1972. Unknown by the American public or the United States Congress, these operations were carried out by the U.S. Army through the Department of Defense under such code names as "Popeye," "Intermediary-Compatriot," and "Nile Blue." From data supplied by the Department of Defense and published by Deborah Shapley in the June 7, 1974, issue of *Science* magazine, Table 1 shows that, in the course of the U.S. Army's cloud seeding operations over southeast Asia, 2602 sorties were flown, and 47,409 cannisters of silver iodide or lead iodide were dropped.

TABLE 1
U.S. CLOUD SEEDING EFFORTS
OVER SOUTHEAST ASIA

YEAR	SORTIES FLOWN	CANNISTERS DROPPED
1967	591	6,570
1968	734	7,420
1969	528	9,457
1970	277	8,312
1971	333	11,288
1972	139	4,362
TOTALS	2,602	47,409

Although the syndicated newspaper columnist Jack Anderson first made the American public aware of these secret military operations in a 1971 column, Deborah Shapley, in further articles published in *Science* magazine, provided details of these once-secret operations. According to Shapley's reports, the seeding cannisters—40-millimeter aluminum photoflash-type cartridge cases with primer and candle assembly—were developed at the Naval Weapons Center in China Lake, California. With a delayed firing mechanism, the cannisters were dropped into clouds, seeding them with either silver iodide or lead iodide. WC-130 weather reconnaissance aircraft and RF-4C reconnaissance aircraft were used in the seeding operations, both of these aircraft being able to hold 104 cannisters each. Each sortie flown was able to seed four to five clouds. While the effect lasted an average of one-half hour, the effects of seeding under ideal conditions could last up to six hours. Each seeding could influence an area 32 kilometers in diameter.

Initially, the area of seeding operations was confined to the Laos panhandle, but later was expanded to include North Vietnam, South Vietnam, and Cambodia. The main target for the operations was the Ho Chi Minh Trail and the movements of enemy troops down it. When President Lyndon Johnson announced a halt to the bombing of North Vietnam above the 19th parallel in March 1968, cloud-seeding operations above the 19th parallel were also suspended. But they were continued in other areas of southeast Asia. The operations were carried out during the southwest monsoon, which generally occurs each year from about April through September. By intensifying the monsoon rains, roads were washed out and made impassable. Calculations by the Defense Intelligence Agency estimate that in some areas of seeding, precipitation was increased by as much as 30%. During the monsoon of 1971, the worst flooding since 1945 hit North Vietnam and destroyed its rice crop. The U.S. Army's cloud-seeding operations are considered to have been a major factor in these floodings.

In light of the fact that global climate conditions tend to follow an east-to-west direction, one must wonder whether the U.S. cloud-seeding operations over southeast Asia could have contributed to the severe drought that struck the Sahel from 1968 to 1973. With a one-year time lag, the timing between the Sahel drought of 1968 to 1973 and the cloud-seeding from 1967 to 1972 seems a strange coincidence, especially if we recall the natural dynamics of weather systems whereby, if we change the weather in one area, we also affect the weather downwind. Could the intensification of the southeast Asia monsoons have depleted the potential precipitation for the Sahel? While we merely raise the question, it is an interesting parallel worthy of consideration before man next attempts to modify the weather, either on such a grand scale or on a local basis.

Weather modifications by the U.S. Army during the Indo-China war were not confined to cloud-seeding. Clouds were also seeded with chemicals that produced an acidic rain effective in disabling North Vietnamese radar units used to control anti-aircraft missiles. The extensive defoliation of southeast Asia conducted during the war effort will have lasting effects on the region. In the final analysis, the war in Indo-China was not one of conflicting ideologies but rather of economics, the control of the region once called the "breadbasket" of Asia. Ironically, defoliation and the spraying of herbicides contributed significantly to the devastation of this previously bountiful land. Arthur Westing of Windham College states that in South Vietnam alone approximately 1.7 million hectares—roughly 10% of South Vietnam—were sprayed one or more times with herbicides during the war. This rich resource of agricultural land has been condemned to wasteland in the immediate future. We also now know that radical climate changes are likely for southeast Asia in the future, with a potential impact on global climate conditions as well.

Nor is southeast Asia alone in suffering the destructive intents of weather modification. It is alleged that in 1970 the C.I.A. carried out cloud-seeding operations to provoke a drought in Cuba and destroy the sugar crop. According to these allegations, the C.I.A. seeded clouds heading toward Cuba and triggered the precipitation of these clouds over the sea. Although one must question the variables involved in the cause and effect, Cuba did experience erratic weather that year, a prolonged dry spell, and a shortfall in the annual sugar harvest, which prompted Premier Fidel Castro to offer his resignation. Lest the reader feel the United States is alone in employing weather modification for negative effects, we should recognize that the flow of information in the United States is greater and more open than in other countries in the world. While the U.S. may have been "caught" using weather modification with malevolent intent, it is quite likely that other countries have carried out, or are presently conducting, weather modification programs for evil purposes that have not been exposed.

The technology of weather modification and its potential for malevolent purposes are terrifying. When, in 1974, the Soviet Ambassador to the United Nations, Jacob A. Malik, introduced a resolution to ban geophysical and environmental warfare, he offered a listing of some of the gruesome means by which scientific advances could be used in military warfare. Recognition of the capabilities to use scientific knowledge in behalf of evil intents led to the December 1974 United Nations Resolution No. 3264 (XXIX) entitled "Prohibition of action to influence the environment and climate for military and other purposes incompatible with the maintenance of international security, human well-being and health."

Resolutions are only as good as the willingness to abide by them. Scientific knowledge and technology advance daily, and with them man's capacity to destroy himself. Instead of dwelling on this ghoulish subject, let us briefly list some of the frightening means man could employ in climatological, environmental, and geophysical warfare:

1. Cloud-seeding to trigger precipitation and cause extensive flooding.
2. Cloud-seeding to trigger precipitation prematurely and prevent rain from reaching areas downwind, perhaps promoting drought conditions in those downwind regions.
3. Cloud-seeding to steer tropical cyclones toward enemy shores.
4. Creating or intensifying fog conditions.
5. Using chemicals in cloud-seeding that produce acidic rain, thereby knocking out radar units and increasing the toxicity in the biosphere.
6. Defoliation of vegetation, reducing lands suitable for cultivation to burned-out wastelands.
7. Adding to the aerosol content of the atmosphere, promoting a cooling trend.
8. The pollution of rivers, increasing the toxicity in the biosphere downstream.
9. The diversion of rivers, cutting off water supplies downstream. Or, as in the case of the Soviet proposal to divert Siberian rivers from flowing into the Arctic in order to irrigate the Siberian farmlands, a diversion resulting in the increase of salinity in the Arctic ocean. This project could promote the melting of the Arctic ice sheets and in turn could cause extensive coastal flooding.
10. Melting the polar ice caps by spreading black soot or particles over the ice cover, reducing the reflectivity of the ice cover. The reduced reflectivity would alter the radiational balance and promote the melting of the polar ice caps.
11. Ice surges. Nuclear detonation at the base of an ice sheet could trigger the sliding of large areas of the ice sheet into the water, creating giant *tsunamis* that would wreak havoc along certain coastal regions. In turn, the ice surge would also change the radiational balance by increasing the reflectivity of surface features, thereby contributing to a cooling trend.
12. Triggering earthquakes. In Chapter VII we saw how man can trigger earthquakes. By adding stress to an

already-strained fault, man could trigger earthquakes further along the fault system.

13. Depletion of the ozone layer in the stratosphere. Through nuclear explosions in the stratosphere or the introduction of ozone-destructive chemicals such as chlorine or nitrogen oxides, punching a hole in the ozone layer over enemy territory would bombard that region with the potentially lethal radiation of ultra-violet.

14. Modifying the chemical character of the oceans, either increasing the salinity of the oceans to promote melting of ice cover or decreasing the salinity to promote expansion of ice cover.

15. Manipulating the electrical properties of the atmosphere, causing weather-pattern changes and interfering with the electrical properties of the human brain, perhaps to the extent of generating depression, anxiety, and mental derangement.

Indeed, these various ways of affecting the climate and environment are a Pandora's Box of man's technological accomplishments gone wild. As climate conditions in the future produce increasingly adverse weather patterns from the natural cooling trend, we will be hard pressed to take matters into our own hands through trying to modify the weather. When the misery of erratic weather affects us directly, we are likely to employ whatever means possible to alleviate our suffering by forcing the hand of nature. The tragedy of the Sahel, like the famine related in the Book of Genesis, could have been averted.

The time to act is now.

It is essential for us to reap the harvests and store a portion as food reserves before it is too late. After the mid-1980's, we are likely to see climate conditions worsen, bringing severe droughts, areas of extensive flooding, and significant shortfalls in food production with which to feed the increasing world population. The scenario for our future may read like doomsday. But it needn't be. Awareness, preparation, and a respect for the power involved in the alternating phases of natural cycles can get us through this most challenging period of change. With a clearer understanding of our territorial limit within this world, we must reevaluate our priorities. We are all citizens of a politically divided, but ecologically interwoven, world. Should we ultimately disable our earth, the awesome task of regulating all her functions would be too much for humanity to handle. If we ruin the earth, there's no place else to go.

CONCLUSION:

THE HEALING OF THE EARTH

"To every thing there is a season, and a time to every purpose under the heaven;

A time to be born, and a time to die; a time to plant, and a time to pluck up that which is planted;

A time to kill, and a time to heal; a time to break down, and a time to build up;

A time to weep, and a time to laugh; a time to mourn, and a time to dance; . . . ."

—Ecclesiastes III:1-4

Here we are in the 1980's. For better or for worse, but needing to do whatever we can to help offset the hazards and to give hope for the emergence of a better world with better people in it. Through the synthesis of intuition and reason, perhaps, we can better understand the conditions we may soon experience. These conditions do not allow for an attitude of complacent security, but give us cause for alarm. Seers and scientists have both confirmed, in their separate ways, that man and the earth are partners in one life-sustaining whole.

They are also in agreement that the future will confront us with radical transformations to an earth we have grown accustomed to. The avenues along which these earth changes will come could be many, and include both an increase in the frequency of volcanoes and earthquakes, as well as a trend towards global cooling. The higher incidence of earthquakes and volcanoes will shake us to the core. Changing weather patterns and climate conditions will make our survival a constant struggle.

We must continue to increase our awareness of the earth as a living, breathing being with its own rhythmic natural cycles. This awareness cannot be confined to the scientific community. It must be disseminated through the general public. From our discussions of geophysical variations in this book, we have gained an understanding of the mechanics involved in volcanic eruptions and earthquake activity. Through the research of the Japanese, Chinese, Russians, and Americans, we know that the concept of dilatancy and the associated changes in various natural phenomena provide warnings of incipient earthquakes. Through the advanced technology of observation satellites and the involvement of the citizenry, similar to the Chinese mobilization after the 1966 Hsingtai earthquake, man can monitor these natural phenomena for anomalous changes. This monitoring will allow for an early warning of potential earthquake or volcanic activity. If people are informed about the effects of anticipated geological variations, they can set up disaster relief teams on a local basis. Emergency supplies can be readied on an individual and community level with provisions of fresh water, food, and medical equipment. If we are prepared to face the coming changes, suffering and destruction can be lessened, tens of thousands of lives can be saved.

We must discard our concepts of "normalcy" and "stability," which make us rigid in our belief that all change is revolutionary and transient rather than an integral factor of life's evolution on this planet. We saw this rigidity in the geological principles of uniformitarianism, stabilism, and the permanency of continents and oceans. These principles stubbornly held on until the discoveries of men such as Alfred Wegener and others found overwhelming evidence against them. When we talk of "normal" weather, we must remember that we are dealing with a concept based on average weather conditions over the past 30 to 40 years—a minute speck of time in the earth's history. Being aware of the mechanics involved in the weather systems is also being able to perceive the dangers involved in our tampering with it. We can no longer rape the earth by transforming its features to satisfy our short-sighted needs. We can no longer agitate the earth with chemicals or other harmful stimulants in order to extort maximal productivity. We can no longer dump our wastes onto the earth without suffering adverse reactions.

Our past obsessions with mastering and controlling our environment by destruction and devastation are outmoded. Those who stubbornly cling to them experience distress, but must face the fact that all changes are unsettling, but necessary if in the end the outcome is to be beneficial.

Too many of us live our lives on the border of meaninglessness. We have so lost touch with reality that we no longer have any idea of the purpose of life. We have focused our efforts on the material aspects of being and have lost sight of the spiritual aspects. Although modern man is mired in an all-consuming material consciousness, it was not always this way. The history of mankind has been a see-saw of duality. Describing man's rapport with nature, Rudolf Steiner speaks of "when a time there was for the consciousness of man no such thing as what *we* call matter . . . Everything in outer Nature was immediately seen as the embodiment of divine-spiritual Beings who manifest themselves throughout the whole of Nature . . . If we look back a few centuries we see that a living relationship existed between man and a divine Being who was living, weaving, and working in natural phenomena . . . ." Call our current state a "fall from grace," if you will, but it's only part of the continuum between the evolution and involution of mankind.

We tend to be so absorbed in materialism that we see the effects, but neglect the underlying causes. We have fed the material aspect of our being and starved the spiritual aspect. The time has come for a choice between short-term profit and long-term survival. We are entering a period of transition filled with anguish and torment in which the securities of stability, fixity, and normalcy must give way before our very life-support systems do. In the face of all this, the tendency toward pessimism, inertia, and downright dejection may seem well-founded.

But this type of reaction—downright reactionary—is neither the only justifiable option nor a terribly healthy one on which to base alternatives.

By reassociating ourselves with the natural dynamics of earth processes, we can reestablish the links of interdependency within the life gestalt. We can reacknowledge our dependency upon the earth for survival, balance our demands with the earth's dictates.

"As a man thinks, so shall he become." Our attitude during this process of rapid change is paramount. Understanding the underlying causes of change and the consequent effects is vital. While greater care and respect in working with the earth is of utmost importance, we must also nurture a renewed respect for ourselves and our fellow humans. This may prove difficult for us. Having fallen into a vortex of negativity, we have tended to dissociate ourselves from the interdependency of the community. The cult of narcissism has become so appealing partly because it eliminates responsibility to others. The competition, greed, and selfishness that are so prevalent in societal behavior have created a high-pitched atmosphere of friction. Resistance and aggression have become the mode of operations in society.

Our culture's sole emphasis on material acquisitions emphasizes the quantity of living rather than its quality. Our appreciation of ourselves and others is not based on *who* we are, but what we have. Our cars, houses, incomes, positions, travels, have become all-important in seeking respect for ourselves. We've devised a material paradise for ourselves that we can barely stand living in. To get through the pains of life, we opt for a nebulous existence, a life shrouded behind a veil of drugs, alcohol, amorality, and self-serving "-ologies" and "-isms." Ignoring a proud heritage and a potentially even better future, mankind seems to have lapsed into indulgent catatonia.

The future will severely test us, demanding maximal strength and extreme resilience on an individual and collective level. As earth conditions continue to worsen, the potential for global conflict will escalate. Challenged by hardship and misery, people are liable to lash out *en masse* in frustration, anger, and violence. In personal terms, in national and international terms, we're responsible for humanity's lot. The world's problems can only be solved from a higher level, a level of global concern, expressed in action for the common good.

In discussing the cause of war, the philosopher George Gurdjieff saw two primary factors. The first is a state of tension that develops periodically within the natural coordinates of the life process. The second is man's reaction to that state of tension. When the world passes through a universal phase of tension, man tends to become dissatisfied with established ways. Dissatisfaction creates tension which is often expressed through conflict, aggression, and war. If people are made aware of the life process, with its alternating phases, the universal reaction to tension might be healthier. Instead of flailing out violently, man could channel that heightened energy level into self-development and the evolution of his inner being.

In light of Gurdjieff's thoughts, it is interesting to discover that utopian literature written in the past has coincided with periods of incredible adverse earth conditions. In the past, humanity has been challenged with energy crises, famine, epidemics, and global conflict. In the past, these challenges have been overcome. A similar response is needed for the future. Honed by the agonies of tomorrow's world, we will gain inner strength, increasing awareness of our spiritual selves, and a greater respect for the life process. If we have willed to develop the material part of ourselves, we can also will to develop the spiritual part, and thus achieve a balance between these two vital aspects of our being. The current insecure quest through competition must be replaced by the loving spirit of cooperation.

One of Buckminster Fuller's recent books captures the essence of the choices before us in its very title: *Utopia or Oblivion*.

It's up to us to promote better ways of living based on entirely different premises of caring, sharing, and community, for our family, our neighborhood, and our world, and in very practical ways, prepare for the changes ahead.

BIBLIOGRAPHY

This bibliography is broken down into sections. Books with material relevant to several sections are cited under the first section only.

SCIENTIFIC AND PROPHETIC VISION: TRANSFORMERS OF THE EARTH

Beveridge, W.I.B. *The Art of Scientific Investigation*. Rev. ed. New York: W. W. Norton & Co., Inc., 1957.

Capra, Fritjof. *The Tao of Physics;* Berkeley: Shambhala, 1975.

THE PROPHECIES OF CHANGE

Bjornstad, James. *Twentieth Century Prophecy*. New York: Pyramid Books, 1970.

Carter, Mary Ellen. *Edgar Cayce on Prophecy*. New York: Warner Books, 1968.

Cerminara, Gina. *Many Mansions*. New York: William Sloane Associates, 1950.

Earth Changes. Virginia Beach, Virginia: A.R.E. Press, 1971.

Earth Changes: Past—Present—Future. Virginia Beach, Virginia: A.R.E. Press, 1959.

Kueshana, Eklal. *The Ultimate Frontier*. Chicago: The Stelle Group, 1963.

Prieditis, Arthur. *The Fate of the Nations*. St. Paul: Llewellyn Publications, 1975.

Robinson, Lytle W. *Edgar Cayce's Story of the Origin and Destiny of Man*. New York: Coward, McCann & Geoghegan, Inc., 1972.

Stearn, Jess. *Edgar Cayce—The Sleeping Prophet*. Garden City, New York: Doubleday & Co., Inc., 1967.

Stearn, Jess. *A Prophet in His Own Country: The Story of the Young Edgar Cayce*. New York: William Morrow & Co. Inc., 1974.

Sugrue, Thomas. *There is a River: The Story of Edgar Cayce*. New York: Henry Holt & Co., 1942.

Woldben, A. *After Nostradamus* (trans. by Gavin Gibbons) London: Neville Spearman, 1973.

ASTROLOGY AFFIRMS

Abbott, R.L. *Astrology-at-a-Glance*. Baltimore: I. & M. Ottenheimer, 1954.

deVore, Nicholas. *Encyclopedia of Astrology*. New York: Philosophical Library, 1947.

Gauquelin, Michel. *Cosmic Influences on Human Behavior*. New York: Stein & Day, 1973.

Goodavage, Joseph F. *Write Your Own Horoscope*. New York: Signet Books, 1975.

The Holy Bible (King James Version). Philadelphia: The John C. Winston Co., 1945.

Lee. Dal. *Dictionary of Astrology*. New York: Constellation International, 1969.

MacNeice, Louis. *Astrology*. Garden City, New York: Doubleday & Co., Inc., 1964.

Mayo, Jeff. *Astrology*. London: Teach Yourself Books, 1964.

McCaffery, Ellen. *Graphic Astrology*. Richmond, Virginia: Macoy Publishing Co., 1952.

Moore, Marcia, and Mark Douglas. *Astrology, the Divine Science*. York Harbor, Maine: Arcane Publications, 1971.

Reid, Vera W. *Towards Aquarius*. New York: Arco Publishing Co., Inc., 1969.

Stearn, Jess. *A Time for Astrology*. New York: Signet Books, 1972.

West, John Anthony, and Jan Gerhard Toonder. *The Case for Astrology*. Baltimore: Penguin Books, Inc., 1973.

THE EARTH HAS A BODY ALSO

Bascom, Willard. *A Hole in the Bottom of the Sea*. Garden City, New York: Doubleday & Co., Inc., 1961.

Dineley, David. *Earth's Voyage Through Time*. New York: Alfred A. Knopf, 1974.

Knopoff, L. "The Upper Mantle of the Earth." *Science*, March 21, 1969, Vol. 163, No. 3873, pp. 1277-1287.

Kuo, J.T., Jachens, R.C., Ewing, M., and G. White. "Transcontinental Tidal Gravity Profile Across the United States." *Science*, May 22, 1970, Vol. 168, No. 3934, pp. 968-971.

Moorbath, Stephen. "The Oldest Rocks and the Growth of Continents." *Scientific American*, March 1977, Vol. 236. No. 3, pp. 92-104.

Orowan, Egon. "The Origin of the Oceanic Ridges." *Scientific American*, Nov. 1969, pp. 102-119.

Pollack, Henry N., and David S. Chapman. "The Flow of Heat From the Earth's Interior." *Scientific American*, August 1977, Vol. 237 No. 2; pp. 60-76.

Purrett, Louise. "Probing the Earth's Inner Core." *Science News*, March 11, 1972, Vol. 101 No. 11, pp. 172-173.

Takahashi, Taro, and William A. Bassett. "The Composition of the Earth's Interior." *Scientific American*, June 1965, Vol. 212 No. 6, pp. 100-106 + .

Takeuchi, H., Uyeda, S., and H. Kanamori. *Debate about the Earth*. San Francisco: Freeman, Cooper & Co., 1967.

FOREVER IN MOTION: OUR MOBILE EARTH

Allard, Gilles O., and Vernon J. Hurst. "Brazil-Gabon Geologic Link Supports Continental Drift." *Science*, Feb. 7, 1969, Vol. 163 No. 3867, pp. 528-532.

"Australarctica." *Scientific American*, Aug. 1969, Vol. 221 No. 2, pp. 50, 56.

Behrman, Daniel. "Hunting Clues to an Ancient Supercontinent." *UNESCO Courier*, July 1970, pp. 28-32.

Calder, Nigel. *Restless Earth*. London: Futura Publications Limited, 1975.

Carrigan, Charles R., and David Gubbins. "The Source of the Earth's Magnetic Field." *Scientific American*, Feb. 1979, Vol. 240 No. 2, pp. 118-130.

Colbert, Edwin H. "Antarctic Fossils and the Reconstruction of Gondwanaland." *Natural History*, Jan. 1972, Vol. LXXXI No. 1, pp. 66-73.

"Continental Collisions: Pangea Revised." *Science News*, June 9, 1979, Vol. 115 No. 23, p. 373.

"A Coriolis Effect for Continents." *Science News*, April 1, 1972, Vol. 101 No. 4, p. 215.

Cox, Allan, Dalrymple, G. Brent, and Richard R. Doett. "Reversals of the Earth's Magnetic Field." *Scientific American*, Feb. 1967, Vol. 216 No. 2, pp. 44-54.

"A Cycle of Plumes for Plate Tectonics." *Science News*, Dec. 16, 1972, Vol. 102 No. 25, p. 391.

Dietz, Robert S. "Those Shifty Continents." *Sea Frontiers*, July-Aug. 1971, pp. 204-212.

Dietz, Robert S., and Walter P. Sproll. "Fit Between Africa and Antarctica: A Continental Drift Reconstruction." *Science*, March 20, 1970, Vol. 167 No. 3925, pp. 1612-1614.

"Drifting Theories Shake Up Geology." *Science News*, April 29, 1967, Vol. 91 No. 17, p. 399.

Elliot, David H., Colbert, Edwin H., Breed, William J., Jensen, James A., and Jon S. Powell. "Triassic Tetrapods from Antarctica: Evidence for Continental Drift." *Science*, Sept. 18, 1970, Vol. 169 No. 3951, pp. 1197-1201.

Frazier, Kendrick. "The Unfathomed Forces Driving Earth's Plates." *Science News*, July 25, 1970, Vol. 98 Nos. 3 & 4, pp. 74-76.

Gould, Stephen Jay. "The Continental Drift Affair." *Natural History*, Feb. 1977, Vol. 86 No. 2, pp. 12-18.

Hallam, A. *A Revolution in the Earth Sciences*. London: Oxford University Press, 1973.

Heirtzler, J.R. "Sea-Floor Spreading." *Scientific American*, Dec. 1968, Vol. 219 No. 6, pp. 60-70.

Hide, Raymond. "Motions of the Earth's Core and Mantle, and Variations of the Main Geomagnetic Field." *Science;* July 7, 1967, Vol. 157 No. 3784, pp. 55-56.

"Hot Spots and Crust Motion." *Science News*, March 13, 1971, Vol. 99 No. 11, p. 180.

Hurley, Patrick M. "The Confirmation of Continental Drift." *Scientific American*, April 1968, Vol. 218 No. 4, pp. 52-62 +

Jordan, Thomas H. "The Deep Structure of the Continents." *Scientific American*, Jan. 1979, Vol. 240 No. 1, pp. 92-107.

Kane, Martin F. "Rotational Inertia of Continents: A Proposed Link Between Polar Wandering and Plate Tectonics." *Science*, March 24, 1972, Vol. 175 No. 4028, pp. 1355-1357.

Knopoff, L., Poehls, K.A., and R.C. Smith. "Drift of Continental Rafts with Asymmetric Heating." *Science*, June 2, 1972, Vol. 176 No. 4038, pp. 1023-1024.

Lessing, Lawrence. "Solving the Riddle of the Shuddering Earth." *Fortune*, Feb. 1965, Vol., LXXI No. 2, pp. 164-168 +.

Marvin, Ursula B. *Continental Drift: The Evolution of a Concept.* Washington, D.C.: Smithsonian Institution Press, 1973.

McElhinny, M.W., and G.R. Luck. "Paleomagnetism and Gondwanaland." *Science*, May 15, 1970, Vol. 168 No. 3933, pp. 830-832.

Menard, H.W. "Sea Floor Spreading, Topography, and the Second Layer." *Science*, August 25, 1967, Vol. 157 No. 3791, pp. 923-924.

Meservey, R. "Topological Inconsistency of Continental Drift on the Present-Sized Earth." *Science*, October 31, 1969, Vol. 166 No. 3905, pp. 609-611.

Ness, Norman F. "Earth's Magnetic Field: A New Look." *Science*, March 4, 1966, Vol. 151 No. 3714, pp. 1041-1052.

Schubert, Gerald, Turcotte, D.L., and E.R. Oxburgh. "Phase Change Instability in the Mantle." *Science*, Sept. 11, 1970, Vol. 169 No. 3950, pp. 1075-1077.

Sclater, John G., and Christopher Tapscott. "The History of the Atlantic." *Scientific American*, June 1979, Vol. 240 No. 6, pp. 156-174.

Shaw, Herbert R. "Earth Tides, Global Heat Flow, and Tectonics." *Science*, May 29, 1970, Vol. 168 No. 3935, pp. 1084-1087.

Sullivan, Walter. *Continents in Motion*. New York: McGraw-Hill Book Co., 1974.

Tarling, D.H., and M.P. *Continental Drift*. England: Penquin Books Ltd., 1972.

"Tracking the Moving Earth." *Science News*, Feb. 14, 1970, Vol. 97 No. 7, pp. 170-171.

Turcotte, Donald L., and E. Ronald Oxburgh. "Continental Drift." *Physics Today*, April 1969, pp. 30-39.

"Under the Spreading Sea Floor." *Scientific American*, Aug. 1967, Vol. 217 No. 2, pp. 40, 44.

Vine, F.J. "Spreading of the Ocean Floor: New Evidence." *Science*, Dec. 16, 1966, Vol. 154 No. 3755, pp. 1405-1415.

"A V-Shaped Clue to the Mantle's Flow." *Science News*, Feb. 5, 1972, Vol. 101 No. 6, p. 87.

Watkins, N.D., and H.G. Goodell. "Geomagnetic Polarity Change and Faunal Extinction in the Southern Ocean." *Science*, May 26, 1967, Vol. 156 No. 3778, pp. 1083-1087.

THE EARTH'S FIERY VOICE: VOLCANOES

Bullard, Fred M. *Volcanoes: In History, In Theory, In Eruption*, Austin, Texas, Univ. of Texas Press, 1962.

Decker, Robert W. "The Anatomy of a Volcano." *1971 Britannica Yearbook of Science and the Future*, pp. 34-49. Chicago. Encyclopaedia Britannica, Inc., 1970.

Francis, Peter. *Volcanoes*. England: Penguin Books Ltd., 1976.

Fuchs, Sir Vivian (ed.). *Forces of Nature*. New York: Holt, Rinehart and Winston, 1977.

Hamilton, Edith, and Huntington Cairns (ed.). *Plato: The Collected Dialogues*. Bolingen Series LXXI. Princeton, N.J.: Princeton Univ. Press, 1961.

Hein, James R., Scholl, David W. and Jacqueline Miller. "Episodes of Aleutian Ridge Explosive Volcanism." *Science*, Jan. 13, 1978, Vol. 199 No. 4325, pp. 137-141.

Heindel, Max. *The Rosicrucian Cosmo-Conception*. Rosicrucian Fellowship, Cal.: Oceanside, 1911.

Krüger, Christoph. *Volcanoes*. New York: G.P. Putnam's Sons, 1971.

"A Link Between Earth Tides and Volcano Eruptions." *Science News*, Oct. 21, 1972, Vol. 102 No. 17, 261.

Post, John D. *The Last Great Subsistence Crisis in the Western World*. Baltimore: The Johns Hopkins Univ. Press, 1977.

Rittmann, A. and L. *Volcanoes*, New York: G.P. Putnam's Sons, 1976.

Schuré, Edouard. *From Sphinx to Christ: An Occult History*. Blauvelt, New York: Rudolf Steiner Publications, 1970.

Sparks, Stephen, and Haraldur Sigurdson. "The Big Blast at Santorini." *Natural History*, April 1978, Vol. LXXXVII No. 4, pp. 70-77.

Stommel, Henry and Elizabeth. "The Year Without A Summer." *Scientific American*, June 1979, Vol. 240 No. 6, pp. 176-186.

"Volcanic Plumes by Plane." *Science News*, April 21, 1979, Vol. 115 No. 16, p. 265.

West, Susan. "The Best-Read Volcano." *Science News*, May 12, 1979, Vol. 115 No. 19, pp. 314-318.

EARTH SPASMS

"Aftershocks Below, Uncertainty Above." *Science News*, Oct. 11, 1969, Vol. 96 No. 15, pp. 322-323.

Aggarwal, Yash P., and Lynn R. Sykes. "Earthquakes, Faults, and Nuclear Power Plants in Southern New York and Northern New Jersey." *Science*, April 28, 1978, Vol. 200 No. 4340, pp. 425-429.

"Alaska Quake Predicted." *The New York Times*, July 10, 1979.

"Alaskan Quake Fills Seismic Gap." *Science News*, March 24, 1979, Vol. 115 No. 12, p. 185.

"Alaskan Quake Potential." *Science News*, June 9, 1979, Vol. 115 No. 23, p. 377.

Anderson, Don L. "Earthquakes and the Rotation of the Earth." *Science*, Oct. 4, 1974, Vol. 186 No. 4158, pp. 49-50.

Behrman, Dan. "China Predicts a Major Earthquake and Saves an Entire Population From Disaster." *UNESCO Courier*, May 1976, pp. 11-13.

Bolt, Bruce A. *Earthquakes: A Primer*, San Francisco: W.H. Freeman And Company, 1978.

Boore, David M. "The Motion of the Ground in Earthquakes." *Scientific American*, Dec. 1977, Vol. 237 No. 6; pp. 69-78.

Brace, W.F., and J.D. Byerlee. "California Earthquakes: Why Only Shallow Focus?" *Science*, June 26, 1970, Vol. 168 No. 3939, pp. 1573-1575.

Brace, W.F., and J.D. Byerlee. "Stick-Slip as a Mechanism for Earthquakes. *Science*, Aug. 26, 1966, Vol. 153 No. 3739, pp. 990-992.

"California's Shifting Crust: Slip Sliding Away." *Science News*, Dec. 17, 1977, Vol. 112 No. 25, p. 404.

Carter, Luther J. "Earthquakes and Nuclear Tests: Playing the Odds on Amchitka." *Science*, Aug. 22, 1969, Vol. 165 No. 3895, pp. 773-776.

"City Under Threat." *The Economist*, Jan. 21, 1978, Vol. 266, No. 7012, pp. 68-71.

Coffman, Jerry L., and Carl A. von Hake (ed.): *Earthquake History of the United States*. Washington, D.C.: U.S. Department of Commerce, 1973.

Crimmin, Eileen. "One Year Later—What Happened in Alaska." *Science Digest*, March 1975, Vol. 57 No. 3, pp. 42-46.

"Dilatancy: A Stamp of Approval." *Science News*, Aug. 19, 1978, Vol. 114 No. 8, p. 121.

Douglas, John H. "Earthquake Research (1): Rethinking Prediction." *Science News*, Feb. 3, 1979, Vol. 115 No. 5, pp. 74-76.

Douglas, John H. "Earthquake Research (2): Averting Disaster." *Science News*, Feb. 10, 1979, Vol. 115 No. 6, pp. 90-92.

Douglas John H. "Waiting for the 'Great Tokai Quake'." *Science News*, April 29, 1978, Vol. 113 No. 17, pp. 282, 283, 286.

"Earthquake and Avalanche. *Science News*, Aug. 1, 1970, Vol. 98 No. 5, pp. 94-95.

"Earthquake: An Evacuation in China, a Warning in California. *Science*, May 7, 1976, Vol. 192 No. 4239, pp. 538-539.

"Earthquake-prediction Studies in China. *Physics Today*, April 1974, Vol. 27 No. 4, p. 19.

Emiliani, Cesare, Harrison, Christopher G.A., and Mary Swanson. "Underground Nuclear Explosions and the Control of Earthquakes." *Science*, Sept. 19, 1969, Vol. 165 No. 3899, pp. 1255-1256

Engdahl, E.R. "Explosion Effects and Earthquakes in the Amchitka Island Region." *Science,* Sept. 24, 1971, Vol. 173 No. 4003, pp. 1232-1235.

"Exploring an Ominous Bulge." *Time,* Jan. 16, 1978, p. 54.

"Forecast of Mexican Quake Accurate, but Ignored." *Science News,* Dec. 9, 1978, Vol. 144 No. 24, pp. 404-405.

Gribbin, John R., and Stephen H. Plagemann. *The Jupiter Effect.* New York: Vintage Books, 1976.

Hagiwari, T., and T. Rikitake. "Japanese Program on Earthquake Prediction." *Science,* Aug. 18, 1967, Vol. 157 No. 3790, pp. 761-768.

Halloran, Richard. "Five Atomic Plants Ordered Shut Down." *The New York Times,* March 14, 1979.

Hamilton, R.M., McKeown, F.A., and J.H. Healy. "Seismic Activity and Faulting Associated with a Large Underground Nuclear Explosion." *Science,* Oct. 31, 1969, Vol. 166 No. 3905, pp. 601-604.

Hammond, Allen L. "Earthquake Prediction and Control." *Science,* July 23, 1971, Vol. 173 No. 3994, p. 316.

Hammond, Allen L. "Earthquake Predictions: Breakthrough in Theoretical Insight?" *Science,* May 25, 1973, Vol. 180 No. 4088, pp. 851-853.

Healy, J.H., and P. Anthony Marshall. "Nuclear Explosions and Distant Earthquakes: A Search for Correlations." *Science,* July 10, 1970, Vol. 169 No. 3941, pp. 176-177.

Healy, J.H., Rubey, W.W., Griggs, D.T., and C.B. Raleigh. "The Denver Earthquakes." *Science,* Sept. 27, 1968, Vol. 161 No. 3848, pp. 1301-1310.

Hill, Gladwin. "Coast Power Company Plagued Since 1958 by Earthquake Faults." *The New York Times,* Dec. 17, 1978.

Hill, Gladwin. "Nuclear Plant Foes Dispute Quake Data." *The New York Times,* Dec. 5, 1978.

Hill, Gladwin. "Study Fails to Pinpoint Cause of Coast Land Bulge." *The New York Times,* July 8, 1979.

Hodgson, John H. *Earthquakes and Earth Structure.* Englewood Cliffs, New Jersey: Prentice-Hall, Inc., 1964.

Houston, Jourdan. "Earthquakes? In New England?" *Blair & Ketchum's Country Journal,* Sept. 1978, Vol. V No. 9, pp. 82-93.

"How to Predict an Earthquake." *Science Digest*, Sept. 1966, Vol. 60 No. 3, pp. 26-27.

"International Views on Quake Prediction." *Science News*, April 28, 1979, Vol. 115 No. 17, p. 279.

" 'Jupiter Effect': Mixed Reaction." *Science News*, Sept. 28, 1974, Vol. 106 No. 13, pp. 197-198.

Kerr, Richard A. "Another Successful Quake Forecast." *Science*, March 16, 1979, Vol. 203 No. 4385, p. 1091.

Kerr, Richard A. "Earthquake Prediction: Mexican Quake Shows One Way to Look for the Big Ones." *Science*, March 2, 1979, Vol. 203 No. 4383, pp. 860-862.

Kisslinger, Carl. "Earthquake Prediction. *Physics Today*, March 1974, Vol. 27 No. 3, pp. 36-42.

Knight, Michael. "Quakes in New England: Common But Mysterious." *The New York Times*, April 19, 1979.

"Magnetic Changes: Clue to Quake Precursor?" *Science News*, Sept. 9, 1972, Vol. 102 No. 11, pp. 164-165.

Mansinha, L., and D.E. Smylie. "Earthquakes and the Earth's Wobble." *Science*, Sept. 13, 1968, Vol. 161 No. 3846, pp. 1127-1129.

McCormick, Sheryl. "A History Lesson in Quakes." *Far Eastern Economic Review*, Aug. 26, 1977, Vol. 97 No. 34, pp. 33-34.

McNally, K.C., Kanamori, H., Pechmann, J.C., and G. Fuis. "Earthquake Swarm Along the San Andreas Fault Near Palmdale, Southern California, 1976 to 1977." *Science*, Sept. 1, 1978, Vol. 201 No. 4358, pp. 814-817.

"Messing with the Mousetrap." *Science News*, Feb. 8, 1969, Vol. 95 No. 6, pp. 138-139.

Molnar, Peter, and Paul Tapponnier. "The Collision Between India and Eurasia." *Scientific American*, April 1977, Vol. 236, No. 4, pp. 30-41.

"New Theories on Quakes." *Science Digest*, Jan. 1975, Vol. 77 No. 1, p. 22.

"NRC Closes Five Nuclear Power Plants." *Science News*, March 24, 1979, Vol. 115 No. 12, p. 184.

"Nuclear Reactors and Eastern Earthquakes." *Science*, March 30, 1979, Vol. 203 No. 4387, p. 1320.

Pakiser, L.C., Eaton, J.P., Healy, J.H., and C.B. Raleigh. "Earthquake Prediction and Control." *Science*, Dec. 19, 1969, Vol. 166 No. 3912, pp. 1467-1474.

Perlman, David. "The Trembling Earth." *1972 Britannica Yearbook of Science and the Future*, pp. 35-49. Chicago: Encyclopedia Britannica, Inc., 1971.

Plafker, George. "Tectonic Deformation Associated with the 1964 Alaska Earthquake." *Science*, June 25, 1965, Vol. 148 No. 3678, pp. 1675-1687.

Press, Frank. "Earthquake Prediction." *Scientific American*, May 1975, Vol. 232 No. 5, pp. 14-23.

Press, Frank, and David Jackson. "Alaskan Earthquake, 27 March 1964: Vertical Extent of Faulting and Elastic Strain Energy Release." *Science*, Feb. 19, 1965, Vol. 147 No. 3660, pp. 867-868.

Purrett, Louise. "The Possibilities of Earthquake Prediction." *Science News*, Feb. 20, 1971, Vol. 99 No. 8, pp. 131-133.

Raleigh, C.B., Healy, J.H., and J.D. Bredehoeft. "An Experiment in Earthquake Control at Rangely, Colorado." *Science*, March 26, 1976, Vol. 191 No. 4233, pp. 1230-1237.

"Reservoir Induced Quakes." *Science News*, Jan. 27, 1979, Vol. 115 No. 4, p. 56.

Richter, Charles F. *Elementary Seismology*. San Francisco: W.H. Freeman & Co., 1958.

"San Andreas Fault Found Shifting at Rapid Rate." *The New York Times*, June 19, 1978, p. 12.

Scholz, Christopher H. "Toward Infallible Earthquake Prediction." *Natural History*, May 1974, Vol. LXXXIII No. 5, pp. 54-59.

Scholz, Christopher H., Sykes, Lynn R., and Yash P. Aggarwal. "Earthquake Prediction: A Physical Basis." *Science*, Aug. 31, 1973, Vol. 181 No. 4102, pp. 803-809.

"Search for the Ultimate Landmark." *Science News*, Sept. 30, 1972, Vol. 102 No. 14, pp. 211-212.

"Shaking Up the Northeast." *Science News*, April 28, 1979, Vol. 115 No. 17, p. 281.

Shaw, Evelyn. "Can Animals Anticipate Earthquakes?" *Natural History*, Nov. 1977, Vol. LXXXVI No. 9, pp. 14-20.

Smylie, D.E., and L. Mansinha. "The Rotation of the Earth." *Scientific American*, Dec. 1971, pp. 80-88.

"Successful Soviet Quake Prediction." *Science News*, March 17, 1979, Vol. 115 No. 11, p. 169.

Sullivan, Walter. "Catastrophic Scope of 1976 China Earthquake is Revealed." *The New York Times*, June 11, 1979.

Sullivan, Walter. "Earth-Squeezing Pushing California South." *The New York Times*, November 26, 1978.

Sullivan, Walter. "New Tools for Predicting Quakes." *The New York Times*, March 27, 1979.

Sullivan, Walter. "Soviet Prediction of Earthquake Proves Accurate in 2 of 3 Factors." *The New York Times*, Dec. 10, 1978.

Sullivan, Walter. "Two Experts Dispute Con Ed Over Quakes." *The New York Times*, April 15, 1978.

"Tangshan Quake: Portrait of a Catastrophe." *Science News*, June 18, 1977, Vol. 111 No. 25, p. 388.

"Tilt Warning on Quake." *Science News*, Jan. 2, 1971, Vol. 99 No. 1, p. 8.

Wakita, Hiroshi. "Water Wells as Possible Indicators of Tectonic Strain." *Science*, Aug. 15, 1975, Vol. 189 No. 4202, pp. 553-555.

"Watch Out, Tokyo and Los Angeles." *The Economist*, Aug. 21, 1976, Vol. 260 No. 6938, pp. 69-70.

"What Keeps the Earth Wobbling." *Scientific American*, Nov. 1968, Vol. 219 No. 5, p. 60.

Whitcomb, James H., Garmany, Jan D., and Don L. Anderson. "Earthquake Prediction: Variation of Seismic Velocities Before the San Fernando Earthquake." *Science*, May 11, 1973, Vol. 180 No. 4086, pp. 632-635.

"The Wobbling Earth." *Science News*, Aug. 14, 1971, Vol. 100 No. 7, p. 108.

BREATHING IN AND BREATHING OUT: THE EARTH'S CLIMATE MACHINE

Abelson, Philip H. "Energy and Climate." *Science*, Sept. 2, 1977, Vol. 197 No. 4307, p. 941.

"Acid Rains and Poison Snows." *The Futurist*, April 1979, Vol. XIII No. 2, pp. 141-151.

Adams, J.A.S., Mantovani, M.S.M., and L.L. Lundell. "Wood Versus Fossil Fuel as a Source of Excess Carbon Dioxide in the Atmosphere: A Preliminary Report." *Science*, April 1, 1977, Vol. 196 No. 4285, pp. 54-56.

Admires, Gerald, and Charles Walters, Jr., "Weather Report." *Acres, U.S.A.*, Sept. 1978, Vol. 9 No. 8, pp. 1, 4-5.

"Air Samples Reveal New Threat to Ozone." *Science News*, Sept. 23, 1978, Vol. 114 No. 13, p. 212.

Alyea, Fred N., Cunnold, Derek M., and Ronald G. Prinn. "Stratospheric Ozone Destruction By Aircraft-Induced Nitrogen Oxides." *Science*, April 11, 1975, Vol. 188 No. 4184, pp. 117-121.

"Antarctic Sea Ice May Herald Ice Age." *Science News*, Jan. 13, 1979, Vol. 115 No. 2, p. 22.

"At War with the Weather." *Science News*, July 9, 1966, Vol. 90 No. 1, pp. 26-27.

Atwater, Marshall A. "Planetary Albedo Changes Due to Aerosols." *Science*, Oct. 2, 1970, Vol. 170 No. 3953, pp. 64-66.

Bandeen, William R., and Stephen P. Maran (ed.). *Possible Relationships Between Solar Activity and Meteorological Phenomena*. A Symposium held at the Goddard Space Flight Center, Greenbelt, Md. November 7-8, 1973. Washington, D.C.: National Aeronautics and Space Administration, 1975.

Barnaby, Frank. "Environmental Warfare." *The Bulletin of the Atomic Scientists*, May 1976, Vol. 32 No. 5, pp. 36-43.

Barnaby, Frank. "Towards Environmental Warfare," *Current*, April 1976, No. 182, pp. 55-59.

Battan, Louis. "Killer Storms." *1972 Britannica Yearbook of Science and the Future*, pp. 114-127. Chicago: Encyclopaedia Britannica, Inc., 1971.

Bickel, Lennard. "Drought Becomes Disaster." *Science News*, Aug. 31, 1968, Vol. 94 No. 9, p. 220.

"A Black Cloud over Cloud Seeders." *Science News*, March 10, 1973, Vol. 103 No. 10, pp. 148-149.

Bolin, Bert. "Changes of Land Biota and Their Importance for the Carbon Cycle." *Science*, May 6, 1977, Vol. 196 No. 4290, pp. 613-615.

Broecker, Wallace S. "Climatic Change: Are We on the Brink of a Pronounced Global Warming?" *Science*, Aug. 8, 1975, Vol. 189 No. 4201, pp. 460-463.

Brown, Lester J. "A Harvest of Neglect: The World's Declining Cropland." *The Futurist*, April 1979, Vol. XIII No. 2, pp. 141-151.

Browne, Malcolm W. "Observers Keep Alert for Storms on the Sun." *The New York Times*, May 19, 1978.

Bryson, Reid A. "A Perspective on Climatic Change." *Science*, May 19, 1974, Vol. 184 No. 4138, pp. 753-760.

Bryson, Reid A., and Thomas J. Murray. *Climates of Hunger*, Madison, Wisconsin: Univ. of Wisconsin Press, 1977.

Butterfield, Fox. "Tree Loss in China Affecting Climate." *The New York Times*, April 18, 1979.

Calder, Nigel. *The Weather Machine*. England: Penguin Books Ltd., 1974.

Callis, Linwood B., Natarajan, Murali, John E. Nealy. "Ozone and Temperature Trends Associated with the 11-Year Solar Cycle." *Science*, June 22, 1979, Vol. 204 No. 4399, pp. 1303-1306.

Carter, Luther J. "Weather Modification: Colorado Heeds Votes in Valley Dispute." *Science*, June 29, 1973, Vol. 180 No. 4093, pp. 1347-1350.

Challinor, R.A. "Variations in the Rate of Rotation of the Earth." *Science*, June 4, 1971, Vol. 172 No. 3987, pp. 1022-1025.

Chameides, William L., and James C.G. Walker. "Stratospheric Ozone: The Possible Effects of Tropospheric-Stratospheric Feedback." *Science*, Dec. 26, 1975, Vol. 190 No. 4221, pp. 1294-1295.

"Changing the Climate." *The Economist*, June 4, 1977, Vol. 263 No. 6979, pp. 88-89.

Changnon, Jr., Stanley A. "Rainfall Changes in Summer Caused by St. Louis." *Science*, July 27, 1979, Vol. 205 No. 4404, pp. 402-404.

Chýlek, Petr, and James A. Coakley, Jr. "Aerosols and Climate." *Science*, Jan. 11, 1974, Vol. 183 No. 4120, pp. 75-77.

"CIA Climate Report: Assessing Impact." *Science News*, May 15, 1976, Vol. 109 No. 20, pp. 310-311.

"Climate for Tomorrow . . . and Yesterday . . . and the Day Before." *Science News*, March 16, 1974, Vol. 105 No. 11, p. 176.

"Clouds Seeded in Coast Storm." *The New York Times*, Feb. 15, 1978.

"Conflicting Reports on Climate Change." *Science News*, Jan. 17, 1976, Vol. 109 No. 3, p. 38.

"Constant Solar Constant?" *Science News*, Aug. 12, 1978, Vol. 114 No. 7, p. 105.

Cooper, Charles F. "What Might Man-Induced Climate Change Mean?" *Foreign Affairs*, April 1978, Vol. 56 No. 3, pp. 500-520.

Cutchis, Pythagoras. "Stratospheric Ozone Depletion and Solar Ultra-violet Radiation on Earth." *Science*, April 5, 1974, Vol. 184 No. 4132, pp. 13-19.

Damon, Paul E., and Steven M. Kunen. "Global Cooling?" *Science*, Aug. 6, 1976, Vol. 193 No. 4252, pp. 447-453.

"Death of the Jungle." *Science News*, Sept. 30, 1978, Vol. 114 No. 14, p. 233.

"Desert Greenhouse." Environment Staff Report. *Environment*, Nov. 1977, Vol. 19 No. 8, pp. 14-20.

Dewey, Edward R., with Og Mandino. *Cycles: The Mysterious Forces That Trigger Events*, New York: Hawthorn Books, Inc., 1971.

Dicke, Robert H. "The Clock Inside the Sun." *New Scientist*, July 5, 1979, Vol. 83 No. 1162, pp. 12-14.

Douglas, John H. "Climate Change: Chilling Possibilities." *Science News*, March 1, 1975, Vol. 107 No. 9, pp. 138-140.

Droessler, Earl G. "Weather Modification." *Science*, Oct. 11, 1968, Vol. 162 No. 3850, pp. 287-288.

"Drought Imperils Sub-Sahara Area." *The New York Times*, Aug. 28, 1977.

"Drought-Solar Link." *Science News*, Aug. 12, 1978, Vol. 114 No. 7, p. 105.

Dubos, René. *A God Within*. New York: Charles Scribner's Sons, 1972.

"Earth Getting Colder as Ice Age Nears." *Science News*, Nov. 26, 1966, Vol. 90 No. 22, p. 449.

Eckholm, Erik, and Lester R. Brown. "The Deserts Are Coming." *The Futurist*, Dec. 1977, Vol. XI No. 6, pp. 361-369.

Eddy, John A. "The Case of the Missing Sunspots." *Scientific American*, May 1977, Vol. 236 No. 5, pp. 80-92.

Emiliani, Cesare "Climate Cycles," *Sea Frontiers*, March-April 1971, Vol. 17 No. 2, pp. 108-120.

Firor, John W., and William W. Kellogg, "Challenging the Restless Atmosphere." *1972 Britannica Yearbook of Science and the Future*, pp. 378-387, Chicago: Encyclopedia Britannica, Inc., 1971.

Fletcher, Joseph O. "Polar Ice and the Global Climate Machine " *Bulletin of the Atomic Scientists*, Dec. 1970, pp. 40-47.

Frazier, Kendrick. "Earth's Cooling Climate." *Science News*, Nov. 15, 1969, Vol. 96 No. 20, pp. 458-459.

Frazier, Kendrick. "The Search for a Way to Suppress Hail." *Science News*, March 20, 1971, Vol. 99 No. 12, pp. 200-202.

Frazier, Kendrick. "The Specter of Meteorological Warfare." *Science News*, July 15, 1972, Vol. 102 No. 3, p. 35.

Frisch, Bruce. "Can We Change the Weather?" *Science Digest*, Sept. 1966, Vol. 60 No. 3, pp. 76-82.

Goody, Richard. "Climate and the Planets." Natural History, Jan. 1978, Vol. LXXXVII No. 1, pp. 84-93.

Gribbin, John. *Forecasts, Famines, and Freezes*. New York: Walker and Company, 1976.

"Hail-Fighting Plan." *Science News*, Sept. 14, 1968, Vol. 94 No. 11, pp. 261-262.

"Hail on the Northern Great Plains." *Science News*, April 18, 1970, Vol. 97 No. 16, p. 384.

Hammond, Allen L. "Weather and Climate Modification: Progress and Problems." *Science*, August 17, 1973, Vol. 181 No. 4100, p. 644.

Hammond, Allen L. "Weather Modification: A Technology Coming of Age." *Science*, May 7, 1971, Vol. 172 No. 3983, pp. 548-549.

Hansen, James E., Wang, Wei-Chyung, and Andrew A. Lacis. "Mount Agung Eruption Provides Test of a Global Climatic Perturbation." *Science*, March 10, 1978, Vol. 199 No. 4333, pp. 1065-1068.

Hanst, Philip L. "Noxious Trace Gases in the Air. Part I. Photochemical Smog." *Chemistry*, Jan./Feb. 1978, Vol. 51 No. 1, pp. 8-15.

Hanst, Philip L. "Noxious Trace Gases in the Air. Part II. Halogenated Pollutants." *Chemistry*, March 1978, Vol. 51 No. 2, pp. 6-12.

Hays, James D. "The Ice Age Cometh." *Saturday Review of the Sciences*, April 1973, Vol. I No. 3, pp. 29-32.

Hays, J.D., Imbrie, John, and N.J. Shackleton. "Variations in the Earth's Orbit: Pacemaker of the Ice Ages." *Science*, Dec. 10, 1976, Vol. 194 No. 4270, pp. 1121-1132.

Hays, James D., and Neil D. Opdyke. "Antarctic Radiolaria, Magnetic Reversals, and Climate Change." *Science*, Nov. 24, 1967, Vol. 158 No. 3804, pp. 1001-1011.

"High Altitude Data Confirm Ozone Theory." *Science News*, August 9, 1975, Vol. 108 No. 6, p. 84.

Hsü, Kenneth J. "When the Black Sea Was Drained." *Scientific American*, May 1978, Vol. 238 No. 5, pp. 52-63.

"Ice Ages Attributed to Orbit Changes." *Science News*, Dec. 4, 1976, Vol. 110 No. 23, p. 356.

"Is the Sun Shrinking? Two Views." *Science News*, June 30, 1979, Vol. 115 No. 26, p. 420.

Kaulins, Andis. "Heliocentric Planetary Positions and their Effect on Sunspots." *Cycles*, June 1977, Vol. 28 No. 4, pp. 88-95.

Kellogg, William W. "Is Mankind Warming the Earth?" *The Bulletin of the Atomic Scientists*, Feb. 1978, Vol. 34 No. 2, pp. 10-19.

Kellogg, W.W., and S.H. Schneider. "Climate Stabilization: For Better or For Worse?" *Science*, Dec. 27, 1974, Vol. 186 No. 4170, pp. 1163-1172.

Kerr, Richard A. "Carbon Dioxide and Climate: Carbon Budget Still Unbalanced." *Science*, Sept. 30, 1977, Vol. 197 No. 4311, pp. 1352-1353.

Kerr, Richard A. "Global Pollution: Is the Arctic Haze Actually Industrial Smog?" *Science*, July 20, 1979, Vol. 205 No. 4403, pp. 290-293.

Kerr, Richard A. "Weather Modification: A Call for Tougher Tests." *Science*, Nov. 24, 1978, Vol. 202 No. 4370, p. 860.

Kukla, George J. and Helena J. Kukla. "Increased Surface Albedo in the Northern Hemisphere." *Science*, Feb. 22, 1974, Vol. 183 No. 4126, pp. 709-714.

Kukla, G.J., and R.K. Matthews. "When Will the Present Interglacial End?" *Science*, Oct. 13, 1972, Vol. 178 No. 4057, pp. 190-191.

Lamb. H.H. *The Changing Climate*. London: Methuen & Co., Ltd., 1966.

Lamb, H.H. *Climate: Present, Past and Future. Volume I: Fundamentals and Climate Now*. London: Methuen & Co., Ltd., 1972.

Lamb, Hubert H. "The Earth's Changing Climate." *1976 Yearbook of Science and the Future*, pp. 180-195. Chicago: Encyclopedia Britannica, Inc., 1975.

Lamb, Hubert H. "Is the Earth's Climate Changing?" *The UNESCO Courier*, Aug.-Sept. 1973, pp. 17-20.

Landsberg, Helmut E. "Man-Made Climatic Changes." *Science*, Dec. 18, 1970, Vol. 170 No. 3964, pp. 1265-1274.

Lansford, Henry. "We're Changing the Weather by Accident." *Science Digest*, Dec. 1973, Vol. 74 No. 6, pp. 18-23.

London, Julius, and Jean Kelley. "Global Trends in Total Atmospheric Ozone." *Science*, May 31, 1974, Vol. 184 No. 4140, pp. 987-989.

Lubkin, Gloria B. "Atmospheric Dust Increase Could Lower Earth's Temperature." *Physics Today*, Oct. 1971, Vol. 24 No. 10, pp. 17, 19-20.

Lubkin, Gloria B. "Do Solar Variations Affect Earth's Weather?" *Physics Today*, Sept. 1975, Vol. 28 No. 9, pp. 19-20.

Lubkin, Gloria B. "Fluorocarbons and the Stratosphere." *Physics Today*, Oct. 1975, Vol. 28 No. 10, pp. 34-39.

MacDonald, Gordon J.F. "How Man Endangers the Climate." *Current*, Jan. 1970, No. 114, pp. 17-24.

"Making the Most of the CO_2 Problem." *Science News*, April 14, 1979, Vol. 115 No. 15, pp. 244-245.

"Making Weather Fit the Crops." *The Futurist*, Feb. 1978, Vol. XII No. 1, pp. 53-54.

Malone, Thomas F. "Weather Modification: Implications of the New Horizons in Research." *Science*, May 19, 1967, Vol. 156 No. 3777, pp. 897-901.

"Man's Impact on Climate: What is Ahead?" *Science News*, July 31, 1971, Vol. 100 No. 5, p. 73.

Maugh III, Thomas H. "The Ozone Layer: The Threat from Aerosol Cans Is Real." *Science*, Oct. 8, 1976, Vol. 194 No. 4261, pp. 170-172.

McConnell, J.C., and H.I. Schiff. "Methyl Chloroform: Impact on Stratospheric Ozone." *Science*, Jan. 13, 1978, Vol. 199 No. 4325, pp. 174-177.

"Measuring Volcanic Particle Production." *Science News*, Sept. 16, 1978, Vol. 114 No. 12, p. 200.

"Military Rainmaking Confirmed by U.S." *Science News*, May 25, 1974, Vol. 105 No. 21, p. 335.

"Military Rainmaking: DOD Still Unresponsive." *Science News*, Aug. 5, 1972, Vol. 102 No. 6, p. 86.

Mueller, Marti. "Hurricane Seeding: A Quest for Data." *Science*, Sept. 5, 1969, Vol. 165 No. 3897, p. 990.

"NAS Warning on Climate Changes." *Science News*, Jan. 25, 1975, Vol. 107 No. 4, pp. 52-53.

Neuberger, Hans, and John Cahir. *Principles of Climatology*. New York: Holt, Rinehart & Winston, Inc., 1969.

Newell, Reginald E. "The Global Circulation of Atmospheric Pollutants." *Scientific American*, Jan. 1971, Vol. 224 No. 1, pp. 32-42.

Newman, James E., and Robert C. Pickett, "World Climates and Food Supply Variations." *Science*, Dec. 6, 1974, Vol. 186 No. 4167, pp. 877-881.

Norwine, Jim. "A Question of Climate." *Environment*, Nov. 1977, Vol. 19 No. 8, pp. 6-13, 25-27.

"Ozone Destruction Exceeds Predictions." *Science News*, Dec. 9, 1978, Vol. 114 No. 24, p. 407.

"Ozone Drop Supports Depletion Theory." *Science News*, Jan. 17, 1976, Vol. 109 No. 3, p. 38.

"Ozone Linked to Sun's UV Flux." *Science News*, June 23, 1979, Vol. 115 No. 25, p. 405.

"Ozone Weathers the Greenhouse Effect." *New Scientist*, July 12, 1979, Vol. 83 No. 1163, p. 87.

Pay, Rex. "Position of Planets Linked to Solar Flare Prediction." *Technology Week*, May 15, 1967, pp. 35-38.

Peterson, James T. "Energy and the Weather." *Environment*, Oct. 1973, Vol. 15 No. 8, pp. 4-9.

"Plant Burning is Major CO Source." *Science News*, March 17, 1979, Vol. 115 No. 11, p. 169.

"Propellants: New Actors in Troposphere?" *Science News*, Oct. 4, 1975, Vol. 108 No. 14, pp. 212-213.

Prospero, Joseph M., and Ruby T. Nees. "Dust Concentration in the Atmosphere of the Equatorial North Atlantic: Possible Relationship to the Sahelian Drought." *Science*, June 10, 1977, Vol. 196 No. 4295, pp. 1196-1198.

Purrett, Louise. "Ice Cores: Clues to Past Climates." *Science News*, Nov. 7, 1970, Vol. 98 No. 19, pp. 369-370.

Purrett, Louise. "The Shifting World of Arctic Sea Ice." *Science News*, July 31, 1971, Vol. 100 No. 5, pp. 80-81.

Purrett, Louise A. "Weather Modification as a Future Weapon." *Science News*, April 15, 1972, Vol. 101 No. 16, pp. 254-255.

"Putting a Check on Ozone Predictions." *Science News*, May 19, 1979, Vol. 115 No. 20, p. 325.

"Rainmaking as a Weapon of War." *Chemistry*, Sept. 1972, Vol. 45 No. 8, p. 6.

Ramanathan, V. "Greenhouse Effect Due to Chlorofluorocarbons: Climatic Implications." *Science*, Oct. 3, 1975, Vol. 190 No. 4209, pp. 50-52.

Rasool, S.I., and S.H. Schneider. "Atmospheric Carbon Dioxide and Aerosols: Effects of Large Increases on Global Climate." *Science*, Sept. 9, 1971, Vol. 173 No. 3992, pp. 138-141.

Reck, Ruth A. "Aerosols and Polar Temperature Changes." *Science*, May 16, 1975, Vol. 188 No. 4189, pp. 728-730.

Reck, Ruth A. "Aerosols in the Atmosphere: Calculation of the Critical Absorption/Backscatter Ratio." *Science*, Dec. 13, 1974, Vol. 186 No. 4168, pp. 1034-1036.

Reck, Ruth A. "Stratospheric Ozone Effects on Temperature." *Science*, May 7, 1976, Vol. 192 No. 4239, pp. 557-559.

Rensberger, Boyce. "14 Million Acres a Year Vanishing as Deserts Spread Around Globe." *The New York Times*, Aug. 28, 1977.

Rensberger, Boyce. "Lag in World Food Output Renews Fear of Famine." *The New York Times*, July 16, 1978.

Report of the Study of Man's Impact on Climate (SMIC). *Inadvertent Climate Modification*. Cambridge, Massachusetts: M.I.T. Press, 1971

Rhinelander, David. "Sunspot Activity Linked to Form of Skin Cancer." *The Hartford Courant*, Oct. 11, 1978.

Roberts, Walter Orr. "Russian Hail-Suppression Experiments." *Science,* June 23, 1967, Vol. 156 No. 3782, p. 1580.

Ruderman, M.A., Foley, H.H., and J.W. Chamberlain. "Eleven-Year Variation in Polar Ozone and Stratospheric-Ion Chemistry." *Science,* May 7, 1976, Vol. 192 No. 4239, pp. 555-557.

Sartwell, Frank. "Abbot: Cycles and Sub-Cycles." *Science News,* Dec. 10, 1966, Vol. 90 No. 24, p. 496.

Schneider, Stephen H., and Clifford Mass. "Volcanic Dust, Sunspots, and Temperature Trends." *Science,* Nov. 21, 1975, Vol. 190 No. 4216, pp. 741-746.

Schneider, Stephen H., with Lynne E. Mesirow. *The Genesis Strategy,* New York: Plenum Press, 1976.

Shapley, Deborah. "DOD Said to Admit to Weather War." *Science,* April 5, 1974, Vol 184 No. 4132, p. 47.

Shapley, Deborah. "Rainmaking: Rumored Use over Laos Alarms Arms Experts, Scientists." *Science,* June 16, 1972, Vol. 176 No. 4040, pp. 1216-1220.

Shapley, Deborah. "Science Officials Bow to Military on Weather Modification." *Science,* Aug. 4, 1972, Vol. 177 No. 4047, p. 411.

Shapley, Deborah. "Senate Bans Use of Weather, Fire as Weapons by DOD." *Science,* Aug. 11, 1972, Vol. 177 No. 4048, p. 499.

Shapley, Deborah. "Weather Warfare: Pentagon Concedes 7-Year Vietnam Effort." *Science,* June 7, 1974, Vol. 184 No. 4141, pp. 1059-1061.

Shapley, Deborah. "Weather Watch." *Science,* Oct. 13, 1972, Vol. 178 No. 4057, pp. 144-145.

"Sharp Curb Ordered on Aerosol Products." *The New York Times,* March 16, 1978.

Shaw, Glenn E. "Properties of the Background Global Aerosol and their Effects on Climate." *Science,* June 25, 1976, Vol. 192 No. 4246, pp. 1334-1336.

"Shifts Change Life." *Science News,* June 18, 1966, Vol. 89 No. 25, p. 495.

Siegenthaler, U., and H. Oeschger. "Predicting Future Atmospheric Carbon Dioxide Levels." *Science,* January 27, 1978, Vol. 199 No. 4327, pp. 388-395.

Singh, H.B., Salas, L.J., Shigeishi, H., and E. Scribner. "Atmospheric Halocarbons, Hydrocarbons, and Sulfur Hexafluoride:

Global Distributions, Sources, and Sinks." *Science,* March 2, 1979, Vol. 203 No. 4383, pp. 899-903.

Smith, Anthony. *The Seasons.* England: Penguin Books Ltd., 1973.

Sofia, S., O'Keefe, J., Lesh, J.R., and A.S. Endal. "Solar Constant: Constraints on Possible Variations Derived from Solar Diameter Measurements." *Science,* June 22, 1979, Vol. 204 No. 4399, pp. 1306-1308.

"Solar Maximum Year Study." *Science News,* Aug. 11, 1979, Vol. 116 No. 6, p. 103.

"Soviets in U.N. Decry Weather Warfare." *Science News,* Nov. 2, 1974, Vol. 106 No. 18, p. 280.

Stoiber, Richard E., and Anders Jepsen. "Sulfur Dioxide Contributions to the Atmosphere by Volcanoes." *Science,* Nov. 9, 1973, Vol. 182 No. 4112, pp. 577-578.

Stuiver, Minze. "Atmospheric Carbon Dioxide and Carbon Reservoir Changes." *Science,* Jan. 20, 1978, Vol. 199 No. 4326, pp. 253-258.

Sullivan, Walter. "Climate Specialists, in Poll, Foresee No Catastrophic Weather Changes in Rest of Century." *The New York Times,* Feb. 18, 1978.

Sullivan, Walter. "Climatologists Are Warned North Pole Might Melt." *The New York Times,* Feb. 14, 1979.

Sullivan, Walter. "How Long Can the Ozone Take the Fluorocarbons?' *The New York Times,* May 15, 1977.

Sullivan, Walter. "Scientists at World Parley Doubt Climate Variations Are Ominous." *The New York Times,* Feb. 16, 1979.

Sullivan Walter. "Some Physicists Believe the Sun is Now Shrinking 0.1% a Century." *The New York Times,* July 6, 1979.

Tannenbaum, Jeffrey A. "Fluorocarbon Battle Expected to Heat Up as the Regulators Move Beyond Aerosols." *The Wall Street Journal,* Jan. 19, 1978.

Tickell, Crispin. *Climatic Change and World Affairs.* Cambridge, Massachusetts: Center for International Affairs, Harvard University, 1977.

"Upsetting the Climatic Balance." *Chemistry,* Oct 1977, Vol. 50 No. 8, pp. 26-27.

Vonnegut, Bernard. "When Will We Change the Weather?" *Natural History,* Dec. 1967, Vol. LXXVI No. 10, pp. 82-89.

Wang, W.C., Yung, Y.L., Lacis, A.A., Mo, T., and J.E. Hansen. "Greenhouse Effects Due to Man-Made Perturbations of Trace Gases." *Science*, Nov. 12, 1976, Vol. 194 No. 4266, pp. 685-690.

Ward, Barbara, and René Dubos. *Only One Earth*. England: Penguin Books Ltd., 1972.

The Weather Conspiracy. Report by the Impact Team. New York: Ballantine Books, 1977.

"Weather Modification: Becoming Respectable." *Science News*, April 11, 1970, Vol. 97 No. 15, pp. 365-366.

"Weatherwar." *Ramparts*, Aug. 1972, Vol. 11 No. 2, pp. 7-8.

Webster, Bayard. "In the Rain Forest, a Complex and Threatened World." *The New York Times*, April 17, 1979.

Weinberg, Alvin M. "Global Effects of Man's Production of Energy." *Science*, Vol. 186 No. 4160, Oct. 18, 1974, p. 205.

"What's Behind This Winter . . . and What's Ahead." *Science News*, Feb. 12, 1977, Vol. 111 No. 7, pp. 100-101.

Whitaker, H.R. "Don't Bet on Weather Modification Yet." *Science Digest*, Sept. 1972, Vol. 72 No. 3, pp. 70-75.

White, Robert M. "Climate at the Millenium." *Environment*, April 1979, Vol. 21 No. 3, pp. 31-33.

Wilcox, Howard A. *Hothouse Earth*. New York: Praeger, 1975.

Wilcox, John M. "Solar and Interplanetary Magnetic Fields." *Science*, April 8, 1966, Vol. 152 No. 3719, pp. 161-166.

"Windy City Weather Effects." *Science News*, Oct. 14, 1978, Vol. 114 No. 16, p. 264.

Wofsy, Steven C., McElroy, Michael B., and Nien Dak Sze. "Freon Consumption: Implications for Atmospheric Ozone." *Science*, Feb. 14, 1975, Vol. 187 No. 4176, pp. 535-537.

"Wood Famine in Developing Nations." *Science News*, Feb. 24, 1979, Vol. 115 No. 8, p. 119.

Woodwell, George M. "The Carbon Dioxide Question." *Scientific American*, Jan. 1978, Vol. 238 No. 1, pp. 34-43.

Woodwell, G.M., Whittaker, R.H., Reiners, W.A., Likens, G.E., Delwiche, C.C., and D.B. Botkin. "The Biota and the World Carbon Budget." *Science*, Jan. 13, 1978, Vol. 199 No. 4325, pp. 141-146.

"World's Cropland Losses Portend Food Crisis." *Science News*, Nov. 4, 1978, Vol. 114 No. 19, p. 308.

Young, Louise B. *Earth's Aura*. New York: Alfred A. Knopf, 1977.

THE HEALING OF THE EARTH

Bennett, J.G. *What Are We Living For?* London: Hodder and Stoughton, 1949.

Steiner, Rudolf. *Mystery Knowledge and Mystery Centres*. London: Rudolf Steiner Press, 1973.

Steiner, Rudolf. *True and False Paths in Spiritual Investigation*. London: Rudolf Steiner Press, 1969.

Printed in the United States
57400LVS00001B/51